# Shaking the Iron Universe

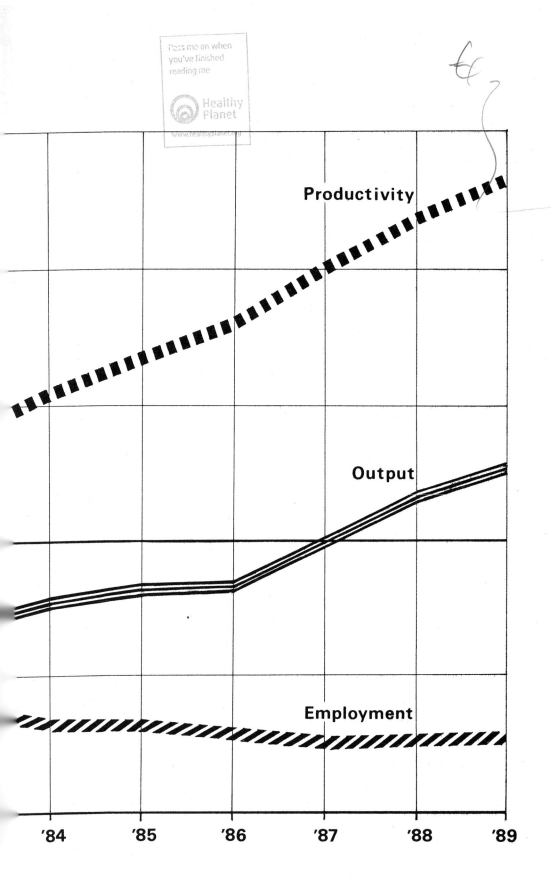

# Shaking the Iron Universe

## British Industry in the 1980s

### David Bowen

Hodder & Stoughton
LONDON   SYDNEY   AUCKLAND   TORONTO

**British Library Cataloguing in Publication Data**

Bowen, David
  Shaking the iron universe.
  1. Great Britain. Manufacturing industries, history
  I. Title
  338.4760941

ISBN 0-340-50847-7

Copyright © David Bowen 1990

First published in Great Britain 1990

All rights reserved. No part of this publication may be reproduced or transmitted in any form or by any means, electronic or mechanical, including photocopying, recording, or any information storage and retrieval system, without either prior permission in writing from the publisher or a licence permitting restricted copying. In the United Kingdom such licences are issued by the Copyright Licensing Agency, 33-34 Alfred Place, London WC1E 7DP. The right of David Bowen to be identified as the author of this work has been asserted by him in accordance with the Copyright, Designs and Patents Act 1988.

Published by Hodder and Stoughton,
a division of Hodder and Stoughton Ltd,
Mill Road, Dunton Green, Sevenoaks, Kent TN13 2YA
Editorial Office: 47 Bedford Square, London WC1B 3DP

Photoset by Butler & Tanner Limited
Printed in Great Britain by Butler & Tanner Limited, Frome and London

# Contents

Introduction vii
Part One  Roots of disaster
 1 Slipping and sliding 3
 2 The sordid Seventies 33
 3 Tory philosophy and the first year 31

Part Two  Depression
 4 Towards the cliff edge 39
 5 Recession 46
 6 Hangover 62
 7 Hiccuping to recovery 78

Part Three  The bull starts to roar
 8 Enter the predators 95
 9 More dealing 110
 10 Death of a giant 120

Part Four  Time for doubt
 11 Better weather 133
 12 Low times for high·tech 142
 13 Laying the charges 154

Part Five  Shaking the iron universe
 14 People upheaval 161
 15 Shake-up 169

## Contents

16 The engineers fight back 184
17 Big bids and politics 200
18 People power 218

Part Six   Rising and falling
19 Crash? What crash? 237
20 Family silver 258
21 Slipping and sliding again 269

Part Seven   Shaken but not stirred?
22 Rebirth in Consett 289
23 Onward and ... upward? 298

Notes 310
Bibliography 312
Index 313

# *Introduction*

In January 1980, Bernard Robinson took over as chief executive of Tallent Engineering at Newton Aycliffe, near Darlington. A local man with great energy, he had come far since joining the company as an apprentice toolmaker in 1956. The same, unhappily, could not be said about Tallent. It had wended its erratic way through the 1950s, 1960s and 1970s, making fancy goods, paraffin heaters, Colston dishwashers, and now parts for the motor industry. It employed 180 people, made sales worth about £3 million a year, and a tiny profit. It was, said Robinson, a "very, very average" British company.

At the end of 1989, Bernard Robinson was still the chief executive of Tallent, it occupied the same factory, and it still made car parts. But that was where the similarities ended. Now, it employed 610 people, had sales of £30 million, and was making a profit of just over £1 million. Scribbles in Japanese on the boardroom blackboard gave a clue as to who Tallent was selling to. Its order book was almost embarrassingly healthy, and the factory was working flat out.

Robinson himself lived in the same semi-detached house as he had ten years before; his lifestyle, he said, was unchanged. (When he was not working, he liked shooting game.) But he had just been voted North East Businessman of the Year, he was now part owner of Tallent thanks to a "management buy-out" and, if he sold his share, he would be a very wealthy man.

This book is, among other things, the story of Tallent Engineering; it tries to explain what Robinson and his colleagues did to transform a small and unsuccessful company into a medium sized and very successful one. It explains how common sense, hard work and an unsentimental willingness to throw away the old and bring in the new (usually copied from someone half way round the world) could transform a company from being a very average British company, which meant a pretty awful one by international standards, to being a good one – by anybody's standards.

It is also the story of other companies, from ICI at the top (in size at least) to tiny PurePlas, which started up in 1988 in Consett, County Durham. Each story can be taken as a piece in a jigsaw, which, when assembled, portrays the fundamental changes that took place in British manufacturing industry in the 1980s.

That, of course, leads to a rather important question: did those fundamental changes ever really happen? Was there indeed a British economic miracle, as many commentators were saying around 1987 and 1988? Or was it all a mirage, as informed opinion was declaring as the 1980s slipped gloomily away? Articles appeared showing that the spurt in productivity had run out of steam, and concluded that British industry was once again heading off down the bumpy slope that was its natural path. "The big rebound in productivity that enabled Britain to start catching up with industry in France and West Germany only seems to have lasted until 1986," the *Financial Times* commented in January 1990. "Its record is now beginning to look tarnished."

That is unnecessarily pessimistic. Although the great leaps in productivity of the first half of the decade were caused mainly by people losing their jobs, and statistics from then on were confused by the great corporate churning stirred up by the City of London, the figures are sufficiently impressive to show that something spectacular must indeed have been going on. Manufacturing productivity (production per person) was 60 per cent higher at the end of the decade than at the beginning; manufacturing employment was 25 per cent lower; output, after slumping, climbed back above the 1979 level in 1987.

More important than all this, though, are the changes in the manufacturing economy that are as yet immeasurable by statistics. They are there, none the less, and by looking at companies like Tallent and the others, and by talking to people who watch and advise them, I have become convinced that there has been a shift in the way that great swathes of industry go about their business. The "iron universe", as G.K. Chesterton called the manufacturing base, has indeed been shaken; probably more than in any previous decade.

What happened, briefly, was this. British industry started the decade with low prestige, a lack of professionalism, confusion about its real

## Introduction

role in the economy, and suffering from the after effects of many years of wildly swinging government policy. Because it had never come under real pressure to change, it had not changed. Subjected to three forces – a profound recession, increased competition from abroad, and the growing threat of takeover – and for the first time faced with a government that was determined not to come to the rescue, it had to turn itself upside down. Much of it did, thanks to new techniques, new managers and, most of all, to the survival instinct. No one would claim that British industry is now the equal of any other – but more of it still exists than could have reasonably been expected in the early 1980s; and some of it is remarkably healthy.

It is difficult to write the history of a particular period even as that period is coming to an end. Apart from the obvious uncertainties – how many of the companies will have gone bankrupt or been taken over before the book is even published? – there is the inevitable fuzziness that comes from looking at something from too close up. Clarity of vision is not helped by the plethora of opinions, many contradictory and all held with the same conviction, that are paraded by economists, politicians and the pundit in the street.

Nevertheless, it is worth trying to get the current state of manufacturing industry into some sort of focus, simply because it is going to be one of the great issues of the 1990s. It was fashionable in the early 1980s for supporters of the Conservative government to assert that manufacturing did not matter. The City, advertising, tourism, and all the other services could, they said, keep 55 million people not only alive but living in ever-increasing luxury on a small and crowded island. That sounded strange to many people then and, as a trade deficit grew up in the last years of the decade, it turned out to be economic nonsense. Britain needs a strong manufacturing base, and there are few who would now deny that.

This book tries to tell the story of the shaking of British industry through the tales of some of those who shook it. By definition, they were successful – and they were not, therefore, a typical cross-section of British industrial managers. If they were, the United Kingdom would probably be suffering from a chronic trade surplus, low inflation and all the other embarrassments that countries like Japan and Germany have to cope with. That said, it is important to make a disclaimer lest any of the companies whose histories I have followed should fall by the wayside in the near future. I have not gone "in search of excellence", looking for companies that will grow ever more robust. Some of the enterprises in this book undoubtedly will do just that, but others that flourished in the 1980s were very much creatures of that decade. The 1990s could provide an environment that is quite hostile to them, or they could simply disappear as part of the continued corporate churning.

But the sample is skewed in other ways – in particular, in the choice of industries. Although most segments of manufacturing are considered, engineering companies, and particularly those in the motor industry, pop up more often than any other. That is because they came closest to extinction in the early 1980s, and made a particularly strong (and largely unremarked) recovery. More important, it is the techniques that they have adopted and adapted that need to spread throughout other parts of industry in the 1990s. Ideas like "total quality management", however cumbersomely named, can be used to streamline operations from the mightiest factory to the humblest workshop. Even the church choir and the local Boy Scout troop could probably, with a little imagination, benefit from them.

Most people who have some interest in business will have followed its progress through newspapers or magazines: jumping from one event to the next, with little guidance as to how one is connected to another. That is the nature of journalism. Many will have followed it from the point of view of their share portfolio, reading the views of journalists who are happy to judge success or failure by the rise or fall of a share price. Much of this book is about the influence of the City: not discussing it would be like writing a history of the Second World War without mentioning the Americans. And it is undeniable that the City was one of the great industrial shakers of the 1980s.

But the City has to be put into context. One stockbroking analyst told me that he was very enthusiastic about firm X. A minute or two later, he said that he thought firm X was one of the worst-run companies in the country. His first comment related to the share price (which he thought was undervalued), the second to its management. It is easy to be confused by a City-centric viewpoint.

The role of other institutions needs to be put into context, too. What was the real effect of the Conservative government? Was it just to make things so bad that industry had to make itself better? Or did it have a more positive role as well? The answer is complex and is, I hope, given in this book. And what of trade unions? Did they have any role at all in the 1980s? Again, that is a simple question with a complex answer. Both questions need to be fitted into the jigsaw puzzle before a reasonably clear picture can emerge.

Talking about industry as "it" brings to mind one homogeneous creature, as though all the manufacturers in Britain became sloppy as one, were shaken as one, and improved as one. That, clearly, is

## INTRODUCTION

nonsense. Different sectors of industry were shaken at different times, and with different force. Furthermore, not every business responded to the shaking as positively as others. For every Bernard Robinson, there is at least one managing director who closed his eyes as the ground began to tremble, and if by good fortune he did not tumble down a crack in the earth, he was free to continue as before when the quake was over. The main problem that British industry had, and still has, is its low status, and the consequent difficulty that it has attracting the best and the brightest to manage it. The process by which the quality of management has improved is another theme – but whether that process can continue, as it needs to, is one of the great concerns of the 1990s.

My personal interest in industry stems, perhaps perversely, from my upbringing in rural Wiltshire. When I was a child, factories were strange and alien things. An education in Oxford and Sussex did little to lessen that feeling, and some of the wonderment remained as I worked as an industrial and financial journalist throughout the 1980s. But as I interviewed people in quite different industries, I kept coming across the same themes and ideas again and again. It is these that I have now tried to gather together.

I could not, however, have done any gathering without the patience and encouragement of my various editors, and particularly of Tony Bush of *Export Direction*, and of Stephen Fay, of *Business*. Having written the book, I was then gently cajoled by Jane Osborn, my editor at Hodder & Stoughton, and Philip Coggan, of the *Financial Times*, until it made a much better read.

Ultimately, the ideas in this book come from the people I have interviewed: managers, management consultants, analysts, and others. It is their comments that appear in quotes, and I would like to thank them all for their help and patience. Finally, I would express my thanks to an under-rated profession – that of public relations. Whatever journalists may say about PR people, they can make our lives a great deal easier.

# Part One

*Roots of disaster*

# I

# *Slipping and sliding*

It would have been astonishing if the Conservatives had not won the General Election in May 1979. Four months earlier, the Winter of Discontent had reached its unsavoury height as rubbish piled up in the streets, unburied corpses had to be stored in a disused factory in Liverpool, and a school janitors' strike kept half a million children at home. Britain had become a laughing stock abroad, and a pretty depressing place to live in.

Mrs Thatcher told the electorate that she stood for change, and they believed her: she was elected with a majority of forty-three on May 3. Her programme for industry, based on a plan drawn up by her mentor and guide, Sir Keith Joseph, sounded simple enough. They wanted to disentangle government from industry; and they wanted a new spirit of "enterprise" to flourish. Simple ideas, but difficult enough to put into practice: 130 years of encrusted anti-industrial attitudes would have to be chiselled away. In the economic sense at least, the Tories really did have to turn the clock back to Victorian times.

The process by which industry slipped and slid to the point where, in the early 1980s, something had to give way was not a complex one. Social attitudes and the whole educational edifice ensured that manufacturing was starved of managerial talent, and the economy was so structured that this simply did not matter. Captive imperial markets and the suppression of competition at home meant that the free market's mechanism to ensure efficiency – competitive pressure – hardly existed. It would have been extraordinary, under these circumstances, if British industry had done anything other than decline.

The last time a major technological exhibition was held in Britain was 1851. The last time a man of industry was hailed as a national hero was in 1859, when George Stephenson's funeral route to Westminster Abbey was lined and the shipyards of the Tyne stopped work in respect. Since then, or at least until the 1980s, the status of the businessman and the engineer has been eroded, and with it the quality of British business.

Martin Wiener, an American historian, has charted the rise of English anti-industrialism.[1] He believes its roots were there from the very beginning. Because the industrial revolution happened first in Britain, it took place at its own pace and without upsetting the established social order. Rather than taking on the rural aristocracy, the new industrialists tended to adapt to its ways and beliefs: the first thing a self-made businessman would do would be to buy an estate. Status became paramount, but to earn it a businessman had to renounce industrialism and, inevitably, lose a grip on the enterprise he had built up. If he did not, his son surely would.

In Germany, by contrast, the new industrialists did clash with the aristocratic junkers and, supported by the Bismarck government – which saw economic growth as the way to political power – it was their ideas that triumphed. In Japan, the Meiji government which took over in 1868 stripped the warrior class of its privileges. In America, of course, there was no rural aristocracy to displace, and industrialists – Carnegie, Ford, Vanderbilt – took on the mantle of aristocracy.

From the middle of the last century, the whole educational and intellectual establishment set about demolishing the self-esteem of British business people. England was seen as two different places: the sordid, smoky "north" and the green and peaceful "south". The image of babbling brooks, market gardens and orchards came to be seen as the *real* England. The Gothic architects and designers such as William Morris consciously looked back to pre-industrial days, while writers like Matthew Arnold, John Stuart Mill and Anthony Trollope made their distaste for the squalid business of industry clear enough.

Public schools, which multiplied in the late nineteenth century, provided a stream of administrators for empire, but were actually producing amateurs with no technical knowledge. "Headmasters more or less equated the classics with civilisation and ideal mental training," Wiener points out.[2] When the state secondary education system was set up in 1902, its schools were modelled closely on the public school ethos. At Oxbridge, "well into the twentieth century, undergraduates were regularly discouraged from pursuing commercial careers, and alarms were sounded against the infection of these rarefied precincts by vulgar influence from without".[3]

Within a few years of Stephenson's death, technical skills had come to be despised in particular. Roy Fedden was sent to Clifton College, a public school, by his sugar merchant father. When, in 1904, he said he wanted to be an engineer rather than go to Sandhurst, there was a terrible family row. Fedden's father "knew his younger son too well to think he had turned 'yellow' and lacked the guts to make a career in uniform. But engineering! It was socially quite unacceptable, being regarded as little better than casual labouring." Fedden went on to become the distinguished chief designer of the Bristol Aircraft Company between the wars.[4]

The contrast with Germany could again hardly be greater. There, a system of technical education was set up that pushed talent from whatever background through the Gymnasium, or grammar school, and on to the local polytechnic. Carl Benz was the son of an engine driver; Gottlieb Daimler's father was a baker. Both went through this system and ended up making, quite independently, the world's first motor cars. In 1899, the Kaiser allowed Prussian polytechnics to award doctorates of engineering – a title still not available in Britain. In 1900, there were 7,130 engineering students in Germany, against 1,433 in Britain.

The great industrial barons had always been people without much formal education. The men who started the industrial revolution were mostly artisans who tinkered with machinery. The mill owners of the nineteenth century were raw entrepreneurs. Where Britain differed from its rivals was that this tradition continued into this century. The early car manufacturers were nearly all bicycle makers who copied what was going on around them, and had a deep distrust of qualifications. "The university mind is a hindrance rather than a help," said motor tycoon Sir Herbert Austin in 1930, while his rival Sir William Morris actively weeded graduates out of his management.[5] In the car industry, virtually no technical breakthroughs were made in Britain before the Second World War.

While the British had an empire that included a quarter of the world's population, that was not important. The imperial trading system meant that foreign competitors posed little threat, however advanced their products.

This century, more and more public school Britons did go into industry. A few, like Fedden, actually wanted to. Owners of companies would also send their sons to paying schools, and in due course the boy would take over as chairman. Other public schoolboys had to go into business because they had no choice: there were not enough "respectable" jobs available. In 1951, a survey showed that 58 per cent of directors of the larger companies had been at public school.

But, with their gentrified attitudes and amateurish detachment from the grubby realities of running factories, they were the "wets" of business. Samuel Courtauld, who inherited chairmanship of his gigantic family firm in 1921, presented the views of those industrialists who saw aims higher than profit. "The quality of workers who leave the factory doors every evening is an even more important thing than the quality of products which it delivers to the customers," he said.[6]

The "dries" were the hardnosed likes of Austin, Morris and Alfred Herbert (who, starting from a bicycle workshop in Coventry at the end of the last century, built up the biggest machine tool company in the world by the 1930s). Another was William Lever (founder of the British half of Unilever), who said in 1909: "There could be no worse friend to labour than the benevolent, philanthropic employer who carries on his business in a loose, lax manner; because, as certain as that man exists ... sooner or later he will be compelled to close."[7]

They were brilliant, autocratic, paternalistic, respected and feared – and invariably they failed to build up a proper management structure. When they went – or lost their touch – "wet" managers would move in and the companies would start to founder. What wets and dries shared, though, was a deep distrust of vocational training, or professionalism.

There was no doubt which way the establishment tended. It was drippingly wet. While Henry Ford was an American folk hero, Morris the industrialist was no more than tolerated (Morris the philanthropist who, as Lord Nuffield, gave away £27 million, was of course admired). Educated Britons found the ruthless pursuit of profit, in the American style, particularly distasteful; and they had a point. For "Taylorism" and "Fordism" were, in their raw form, pretty unpleasant.

Frederick "Speedy" Taylor was, by all accounts, a strange fellow, obsessed with detail and rigid rules. He became at the beginning of this century the chief proponent of "scientific management". Work-study engineers, armed with stopwatches, would calculate exactly how long it should take for a task to be done, and make sure that that was the time allowed. Henry Ford took the system up with gusto. His Highland Park plant in Detroit, which started in 1913, had a continuously moving assembly line, so that cars could be bolted together with maximum efficiency, and according to a timed schedule. Ford combined this regime with brutal suppression of unions to the point that, in 1928, the *New York Times* called him "an industrial fascist, the Mussolini of Detroit".[8] It was hardly surprising that the British clung while they could to their "higher ideals".

Governments connived to keep unbridled competition firmly on the other side of the Atlantic. It was not a political matter: the Tory Party was deeply rooted in the anti-industrial "babbling brook" ethos, while socialists abhorred the dehumanisation of Fordism. Stanley Baldwin, Conservative prime minister in the 1930s and inheritor of an ironworks business, was scandalised by raw capitalism. "Laissez-faire," he said in 1935, "is as dead as the slave trade".[9] In 1943, Lord Hinchingbroke said that "true Conservative opinion is horrified at the damage done to this country since the last war by 'individualist' businessmen, financiers and speculators ranging free in a laissez-faire economy".[10] After the Depression of the early 1930s Baldwin introduced partnership with industry – his government subsidised unprofitable firms – and this partnership grew naturally during the controlled years of the Second World War.

Two industries, chemicals and pharmaceuticals, never suffered from the same stigma as engineering. That was because they were based on pure science, as practised in the great laboratories of Cambridge, fount of so many Nobel prizes. Their strength was based on cleverness, much admired in Britain, rather than competence, the underrated skill of the engineer.

While these sectors were technically successful, they too suffered from poor management which failed to commercialise new ideas efficiently. In part, this was due to the deliberate suppression of competition, particularly by the formation of Imperial Chemical Industries in 1926. It was created out of four chemical companies, Brunner Mond, Nobel Industries, United Alkali and British Dyestuffs Corporation, specifically to provide a British force that would balance the power of Dupont of the USA and IG Farben of Germany. The merger was seen as a way to stop unnecessary price wars between the British groups.

It was not until the 1950s that restriction of competition was seen as conceivably detrimental. Cartels and price-fixing agreements flourished, especially after the Depression, with the law – when tested – tending to decide that they protected jobs and wages.

Had it not been for the Second World War, British business could have continued as before: inefficiently but paternalistically, with the ravages of the free market blunted in a civilised, English manner. New products and new markets might be investigated, but not too rigorously. The Empire would absorb British products uncomplainingly, while the home market could be defended by tariffs; the motor industry was protected by a 33 per cent import tax from 1915; and sweeping protection was imposed in 1931. The British share of world export trade in manufactured goods fell, from 27.5 per cent in 1911–13 to 18.5 per cent in 1931–8 – but that was plenty to keep the economy ticking over.

Trade unionist pressures flared up in the heady years following the First World War, but collapsed in the General Strike of 1926 and remained muted through the depressed 1930s. The degree of unionisation went from 12.4 per cent in 1901, up to 33.1 per cent by 1924, then down to 25 per cent in 1935. In the 1930s, car companies were hardly unionised at all. It was only during the war, when production stoppages had to be avoided at all cost, that unions were allowed in.

The social upheaval triggered by the war was not matched in Britain by a change in business attitudes. In other countries, it was. The Japanese and Germans had no choice but to give even greater precedence to those who could rebuild their country, while the French set up a planning system – with elite engineers, civil servants and managers being pumped out of the Grandes Ecoles to reconstruct the power of France together. The Americans, despite their efficiency, kept to their own patch and never became trading rivals of the British. The reconstructing European countries, along with Japan, had no such inhibitions.

For a few years after the war, it looked as though British industry did have the bit between its teeth. Production in most areas rocketed and, exhorted by the government to export to fill the "dollar gap", it responded. In 1951, when the new spirit of optimism was reflected in the Festival of Britain, 77 per cent of all output was sold abroad.

But it was all an illusion, based on great advantages that were soon thrown away. Although technological strides had been made during the war, especially in electronics, man-made fibres and plastics, and investment had been poured into engineering factories, little had been done to narrow the gap in productivity that had become increasingly apparent in the tariff-protected 1930s. As the historian Corelli Barnett says in his *Audit of War*, "because the Spitfire certainly equalled the Messerschmidt BF 109 in overall performance as a fighter aircraft, it does not follow that Vickers-Supermarine could have sold it successfully in commercial competition ... since the Spitfire took two-thirds more man hours to build".[11] The problems were familiar ones associated with unprofessional management, outdated equipment, factories that were too small, lack of planning... They were exacerbated during the war by union problems, as workers in the shipyards and other industries realised their power. As experts sifted through the wreckage of German industry after the war, they realised that its efficiency, if reconstructed, would soon pose a competitive threat.

But, although Britain was kept from bankruptcy only by American aid, political momentum was directed entirely towards social reform. The Beveridge Report, which laid the basis for the Welfare state, had been drawn up in 1942, and by the time Attlee came to power in 1945, the priorities were health, housing, education, and full employment. Any number of studies were produced predicting that British industry was in no fit state to take on revived competitors, but political imperatives dictated that the few resources there were should be directed towards social projects.

The government tackled the short-term balance of payments crisis by exhorting companies to export – and backed this up by channelling raw materials to those that sold most abroad. Inevitably, managers thought naturally in terms of the Empire rather than the industrialised world. In 1945, it was fairly clear which economies were in better shape, and a combination of tariff and culture gave the British a clear advantage. But the attitude of avoiding competition where possible was deep-rooted. Donald Stokes, later architect of British Leyland, was a Lieutenant-Colonel in Italy in 1945 when he was asked by his old boss Henry Spurrier, chairman of Leyland Motors, to write a brief on the way the firm should organise its export business when the war was over. He wrote that European countries with large sterling balances would represent stop-gap markets before they re-established their own industries. "It was therefore the old Imperial markets, the Middle East and South America, on which Leyland must concentrate."[12]

In 1946–8, the UK made 66 per cent of all the cars in Europe; they were exported all over the world. But models were not rationalised, no attempt was made to adapt them to foreign markets, proper service networks were not set up and, with the exception of sports and luxury cars, they soon became the target of ridicule.

The post-war Attlee government's industrial strategy, such as it was, was halfheartedly interventionist. It nationalised a number of industries, but stopped well short of the full-scale economic planning of the French. When Sir Stafford Cripps, the Labour president of the Board of Trade, exhorted car manufacturers to get together to produce a British "Volkswagen" or standard car, 50 per cent of the output of which should be exported, he was heckled and ignored. In France, the whole thing would have been fixed up over a quiet glass of champagne.

As managers returned to their companies from war duty, they were in no mood to push for upheaval. (Their bosses, of course, had remained in place throughout the war because of their age.) Nor did that matter much; for fifteen years after the war ended, demand outstripped supply. There was no need for companies to strive for efficiency, and they did not. World trade was growing so fast that it mattered little that Britain's share of it fell from 21.4 per cent in 1951–3 to 16.5 per cent in 1959–61: absolute output rose.

The motor car was one of the most important drivers of economic growth, and indeed the motor industry came to symbolise the strengths and weaknesses of the industrial economies. British car production went from 476,000 in 1950 to 1.6 million in 1966, but the supply of cars did not catch up with demand until the late 1950s. Yet burgeoning inefficiency was shown in productivity figures. In 1950, the British produced 3.3 vehicles per employee year, the Americans 10.0 and the Germans 2.2. By the late 1950s, the British had slipped to the bottom of the productivity table and in 1965, the figures were 5.8, 13.9 and 7.1.

As the Empire was dismantled in the first twenty years after the war, newly efficient rivals started to peck away at Britain's not-so-captive markets. British managers did not, on the whole, rise to the challenge. Technical leads were lost through inadequate investment and failure to adopt new technology. The chemical industry grew fast in the 1950s – but not as fast as its European rivals. Shipyards failed to modernise. In 1950, Britain was the leading exporter of ships, selling 38 per cent of the world total. By 1956, she had been overtaken by Japan and Germany, and her share was down to 14 per cent.

Britain's technical expertise was given a boost by the war, but business failed repeatedly to translate it into profits and jobs. The links between pure research and commercial reality stayed further apart than in any other country. The exceptions were mostly in areas where the government provided the impetus as main customer. The aviation industry received constant infusions of state money – as a result, the first jet liner to fly was the De Havilland Comet. After three crashes in 1954, later traced to metal fatigue, the Comet was withdrawn. The Boeing 707 took up the baton, and gave the Americans a lead they never lost.

The first indicators of trouble brewing arrived even before Harold Macmillan told the electorate it had never had it so good, in 1959. Before the war, Imperial Typewriter was a substantial company that exported 40 per cent of its production. In the 1950s it expanded, building another factory, but by the end of the decade, it was already struggling against US and European rivals. Imperial's factories, bought by the American Litton Industries in 1966, were in due course closed down. Labour relations had declined but, said a study, the main problem was "a lack of foresight in the development of new products and the generally relaxed style in which the company was managed". Not only was the percentage of British national income spent on investment consistently lower than that of other countries, it was also less effective: machines tended to be used to cut costs rather than to allow more sophisticated products to be manufactured. Thus (and this was a trend that continued), British companies found they were competing more and more on price, and less on technical skill. Their profit margins were eroded, and thus their ability to invest. It was a vicious circle.

Government attitude to industry was confused. On the one hand, there was a gradual move to create more competition. After the toothless Monopolies and Restrictive Practices Act of 1948 came the formation of the Restrictive Practices Court in 1956. It forced 1,500 restrictive agreements to be abandoned or altered. Then, in 1965, the Monopolies and Mergers Act was passed, allowing government to block takeovers.

On the other hand, neither Labour nor Conservative governments felt any restraint about fiddling around with market forces. From 1947, regional policy forced companies to expand in high unemployment areas, rather than extend their existing plants. For car companies this was particularly serious as it deprived them of much-needed economies of scale. In the early 1960s, Ford and Triumph set up factories on Merseyside, while Rootes built a plant in Linwood in Scotland. The motor industry suffered too from its unchosen role as an instrument of government policy. Because so many companies were dependent on motor output, a convenient way of speeding up or slowing down the economy was to adjust demand for motor cars by changing purchase tax or hire purchase regulations.

While foreign car makers expanded output steadily, pulling huge chunks of domestic industry along with them, British manufacturers held back on investment, while industrial relations fermented as pressures were switched on and off. When the knob was turned to on, Lucas chairman Tony Gill remembers, "the industry had to work balls out to feed demand, and give in to workers who demanded more money". Not surprisingly, the workers took a different view. "They treat labour as machinery here," said one Ford Halewood shop steward. "In many cases they are *cruel*. All they're concerned with is production problems. It's production, production all the time with them."[13] One study at the time suggested that plant managers often welcomed, or even provoked, strikes which could get them off the hook of a missed output target.

The steel industry, too, suffered from changes in policy. Steel companies were used to government involvement – they first started working together in the 1930s to counter the Depression. But they became a political football: nationalised in 1951, gradually privatised over the next ten years, renationalised in 1967, they were damaged by incessant government tinkering. Even today, BSC suffers from Harold Macmillan's 1959 decision to give Wales and Scotland a new strip mill each, when commercially there should have been only one. Although overall output was pushed up in the 1950s, Britain's share of world steel output fell from 10 per cent in 1950 to 5 per cent in 1966, with output per man a third of the EEC or US average.

By the early 1960s, as the last major colonies were given their independence, the warning signals of industrial decline were impossible to ignore. But in 1964, Harold Wilson was elected prime minister and, as the Beatles toured America, many Britons felt the country was rising again in a new and exciting form. Wilson tried to spread this new mood to industry: his catchy promise that Britain would benefit from the "white heat of a new technological revolution" suggested that finally it was going to get its prestige back. The revolution's symbols were Concorde and the Advanced Gas Cooled Reactor: wonderful for morale, but not the basis of a sensible industrial strategy. Government research money was pumped into prestige projects, and into defence, while the Japanese were quietly concentrating on getting their car and engineering industries right.

In retrospect, the most positive move was probably the upgrading of Victorian advanced technology colleges to universities, the rapid expansion of the redbricks, and the establishment of polytechnics to pump out a new breed of pro-technical Briton. It was these creatures that came to the fore twenty years later.

At the time, much more fuss was made of the eight new "plateglass" universities, which had been planned under the Macmillan government. They were set in the countryside, were strong on social sciences and were infused with an anti-industrial ethic. With the exception of Warwick, a piston's throw from Coventry, their scientific aspirations were theoretical, not applied. And the graduates they turned out were, if anything, even less in touch with the real world than those from Oxbridge.

Corporatism was the new word of the 1960s: it meant controlling the economy through the combined efforts of government, business and unions. It was embodied in the Confederation of British Industries, set up in 1966; in the National Economic Development Council (Neddy), where business, unions and government could meet; and in the Industrial Reorganisation Corporation, Labour's grand attempt at saving British industry. The IRC was established in 1966 to use "every available means to improve the competitiveness of British industry". Usually, this meant creating bigger, and theoretically better, groups to take on the world.

All it was doing was reinforcing industry's own reaction to falling competitiveness – which was to form mergers. Between 1960 and 1969, there were 5,635 mergers, against 1,867 in the previous decade. Of the 100 largest manufacturers in 1948, twenty-seven had been taken over by 1968. Many of the acquisitions were to try to increase monopoly power: ICI's unsuccessful bid for Courtaulds in 1961 comes into this category. Others were acquisition for the sake of it – motivated by a simple desire for size by the directors – or because of a vague feeling that companies should be diversifying. For example, Imperial Tobacco bought up a basketful of food companies as well as Courage the brewers during the 1960s.

Studies of this wave of mergers showed that the earnings of merged groups almost invariably fell, and rationalisation was rarely carried through. Arnold Weinstock's 1967 merger of AEI and English Electric with GEC was a brilliant exception. Other managers lacked his skill: they were following the very 1960s belief that big was beautiful. The unions went along with the idea happily enough: they found it easier to negotiate with big firms, and thought they gave better job security. In 1966, when England's prestige was raised to new heights by victory in the World Cup, more people than ever before or since worked in British manufacturing industry: 8.4 million.

The government gave the takeover bandwagon several shoves. After the Geddes report on the shipbuilding industry, which criticised poor equipment, sites, demarcation disputes and high costs, the government contributed £68 million to amalgamate sixty-two yards into three or four large combines. It made little difference. Labour relations remained terrible, and many yards staggered on only with the help of government subsidy.

The IRC, which was headed by the great corporatist industrialist and chairman of Courtaulds, Sir Frank Kearton, was also involved in the concentration of the aircraft industry, the computer industry (into ICL), the ballbearing industry (RHP), the electronics industry and, of course, with the formation of the British Leyland Motor Corporation.

The British Motor Corporation, formed in 1952 when Austin

merged with Morris, was then the biggest car company outside America. It failed to rationalise its production facilities and, by 1959, all sixteen Morris factories (and Austin's one) were still operating. *The Economist* noted then that "there can be few industrial organisations which have more labour troubles than BMC. It is short of good personnel to whom the top management could leave the handling of negotiations."[14] Its two great successes – the Morris Minor and the Mini – were the work of one genius, Alec Issigonis. Yet the extraordinary lack of professionalism was shown in the Mini's pricing: from its launch in 1959, it was sold at a loss simply because BMC did not know (until Ford told it) how much the car cost to build.

Leyland Motors, by contrast, had done well. Based on a Lancashire lorry and bus operation, it had absorbed a handful of other commercial vehicle companies in the 1950s, and Standard-Triumph in 1961. Production of the car company had been rationalised and several successful models had been introduced. It was a company driven by sales, with Donald Stokes storming around the world selling his extensive range of lorries. He took charge in 1963, and became a founder member of the IRC. Like most top industrialists, he liked to work closely with the government, and was unable to resist when his friend Wilson pressed him to take over BMC. Almost immediately, he regretted it.

Feeding, and feeding off, merger mania was a new sort of financial operator. Jim Slater was the best-known. Trained as an accountant, he rose to become deputy sales director of Leyland Motors in the early 1960s, with a good chance of rising to the top. But his real interest was finance, and in 1963 he started contributing a share tipping column to the *Sunday Telegraph*, where Nigel Lawson was the City editor: often he would buy ahead of a tip, or sell after it. The next year, with Tory MP Peter Walker as little more than a sleeping partner, he set up Slater, Walker as an investment service. Soon, he started buying companies up, and selling off their more valuable assets, a process that later became known as asset stripping but was seen then as a hardnosed way of bucking complacent managers up. The City loved Slater, and backed him ever more enthusiastically as his company's market valuation went from £4 million in 1966 to £135 million in 1969. He did not, however, do much to improve the strength of the companies he bought: he was a dealmaker, not a manager.

By the time the fourteen major steel companies were renationalised in 1967, they were in a terrible state. Their technology had been made obsolete by the Japanese, who showed that economies of scale were massive, especially if basic oxygen converters were used instead of traditional open-hearth furnaces. BSC had a dozen open-hearth furnaces, and no docking facilities for ships of more than 20,000 tonnes – when new techniques required ore to be brought in in giant carriers. The new BSC board, led first by Lord Melchett and then by Sir Monty Finniston, developed a rationalisation plan, under which steelmaking would be concentrated at five coastal sites. Progress on rationalising and cutting was started, but soon became bogged down in the industrial quagmire of Britain in the 1970s.

In this climate, the lack of commercial spirit was naturally reinforced. New factories could only be put up after horse-trading with the government, so most industrialists became as chummy as they could with Westminster and Whitehall.

Managers in the very largest companies even started to pretend they were not businessmen at all. "We are very similar," said an ICI executive in the late 1960s, "to the Administrative Class of the Civil Service." They were "enlightened capitalists": ICI was one of the first companies to introduce paid holiday, and was always in the forefront of social advances. But they saw their role as preserving the status quo – and that excluded them from the single-minded pursuit of profit. They fell in naturally and happily with the government-sponsored belief that their role was the preservation of jobs, not the creation of profits. A club atmosphere permeated the top layers of industry, with a corporate class system reinforced by status symbolism. Even GEC, a modern and thrusting company, had nine levels of dining room.

The split between "wet" and "dry" businessmen was as strong as ever, though both shared a love of the status quo. Public school educated wets were found at the top of large companies, or running businesses that they had inherited. The dries, engineers but never graduates, were running their own metal-bashing businesses. Gareth Davies, now chairman of Glynwed International, started work at Steel Parts, a Glynwed subsidiary, in 1957. His managing director was a typical dry: "He was an intelligent guy, a workaholic. He knew the business inside out and knew daily whether he was making money. But managers in those days didn't learn to delegate, and resisted external change." Davies, who was an accountant, thought himself lucky to get a job in industry: experience, not qualifications, were what counted and standardised accountancy was not much regarded. Accounting systems were often esoteric, and understood by no one outside the company. "Financial controls, such as they were, were to keep the group tidy," he says.

At all levels of industry, a sense of cosiness still pervaded. Ian Hays (now a shop steward at Leyland Daf) joined Leyland Motors as an apprentice in 1957. "My first day at work, I got on to the W41 works bus with my pressed overalls and clogs, and the guy who collected the fares said, 'Right lads, this is Ian Hays, and that will be his seat and that's where he puts his butty box, and he'll be here for the duration.' To me, that meant 'til I retired." The Leyland workers could, his colleague Barry Morris adds, "sit in this little corner town in Lancashire and pretend we could take on the world".

It was during the 1960s that trade union power reasserted itself. With the union-financed Labour government in power, this was hardly surprising. In 1968, the power of the TUC was demonstrated when Barbara Castle's *In Place of Strife*, by which some union privileges would be strengthened in exchange for a reduction in the freedom to strike, was defeated. Too often, unions were badly led and too often, bad leaders were unquestioningly followed.

The reaction of most large companies to growing trade union pressure was to build up their own bureaucracies to cope with it. Personnel departments, which had grown fat since the war, were joined by special industrial relations departments. They were experts at the great British compromise – which inevitably excluded radical change. As top management came to accept industrial disruption as part of life, it also found itself insulated by layers of executive and trade union bureaucracy from the workers on the shop floor. The status of the status quo was never higher; and "them-and-us" attitudes were never more entrenched.

In the late 1960s, too, the student revolution took place. Warwick undergraduates protested against their links with industry. The hippy culture was anti-materialist, anti-industrial – similar, in some ways, to the culture of the British establishment.

And yet it was in the 1960s that the seeds of Thatcherism were sown. Attitudes were starting to change. Graduates came into industry out of the new technical universities uncluttered by visions of pre-industrial England and seeing clearly the way that the economy was heading. Typically they came from modest backgrounds and had been to grammar school. They rejected the old ways. "I remember at AP Leamington Spa being told that it would take me thirty years to become as competent as the current middle managers," says Chris Burnham, now director of the consultancy Ingersoll Engineers. "We were all angry young men, getting the same messages."

The influx of foreign-owned firms brought new attitudes and practices with them. By 1970, sales per employee of multinationals in Britain were almost twice as high as among all firms. Inevitably, as managers moved between foreign and British-owned firms, some of these practices started to spread.

Ford, for all its labour troubles, was among the first firms to professionalise management. Its graduate recruitment scheme was started in 1948, and a graduate training scheme began in the mid 1950s. The finance department was shifted from a purely recording to an analytical role in 1956, giving Ford the unusual ability to discover how much its products cost to make. The first fully costed car was the Cortina, which started production in 1962 and was produced at sixteen shillings (80p) under budget.

By the end of the 1960s, some British firms were getting the message. It arrived first in the marketing departments, where men like Ernest Saunders (later chief executive of Guinness) became fascinated by the intriguing theories of market segmentation imported by the Americans. Advertising, marketing's glamorous sister, started to attract some of Britain's brightest graduates. The accountant, while hardly glamorous, also began to play a more important role. He was derided by the engineer as a "bean counter", but was quickly able to prove that figures were really quite a useful way of monitoring a company's performance. The first "numbers-based" conglomerates appeared: Weinstock's GEC, Hanson Trust, which started up in 1969, and BTR, which was grabbed by a team of accountants in 1966 and given a thorough shaking up.

In 1970, the year that John Ashcroft (who became head of Coloroll) joined Tube Investment's graduate training scheme, the firm decided that managers should not be allowed to use their military titles. "One director was a colonel," says Ashcroft. "It was a hell of a traumatic thing when the maintenance guy arrived and changed his name plate to Mr." But that did not change his attitude. "The salesmen in London used to have to go and pick up his coffee from Fortnum & Mason and collect his shirts from Harrods."

This was a time when new was clashing with old. Although TI had a large graduate recruitment programme, most graduates left after a few years – they just did not fit into the firm's culture. American management techniques were examined, and sometimes adopted: it became fashionable to bring in management consultants, with McKinsey's preaching the benefits of matrix management, an elaborate multi-dimensional organisation structure. The journal *Management Today* was launched to keep the modern executive on top of the latest developments, and soon became a bible of change. But the same, old-fashioned type of people were still in charge, and amateurism still ruled. As the old aristocracy had blunted the edge of Victorian entrepreneurs, so the old type of businessman was still in the ascendant. It would take a violent shock to clear it out.

Just as important in the long run was the subtle change in entrepreneurial attitude that took place in the 1960s. In the middle of the decade, the rate of new firm formation started to rise sharply and the total number of small firms in the economy also increased, following a long-term decline.[15] Until 1968, the share of the 100 largest firms in total manufacturing output had been rising steadily: it then hit a plateau. Smaller firms, which are almost by definition more dynamic, were playing a greater role in the economy: this, probably, marked the beginning of the move towards individualism with which Thatcherism is now identified. But before its benefits could find their way into the economy, there was to be a hiatus – the 1970s.

2

# *The sordid Seventies*

In 1970, the Beatles were dissolved, the Vietnam War was at its peak, and the Conservative Edward Heath confounded the polls by winning the General Election. After six years of Wilsonian corporatism, Heath started out determined to give businessmen their head. In 1969, he described Britain as "a Luddite's paradise ... a society dedicated to the prevention of progress and the preservation of the status quo". Four years later, he went to the heart of the matter. "The alternative to expansion is not, as some occasionally seem to suppose, an England of quiet market towns linked only by trains puffing slowly and peacefully through green meadows. The alternative is slums, dangerous roads, old factories, cramped schools, stunted lives."[1] But by then, he had made his attempt at a business-led revolution, and had failed.

The 1970s was a singularly unsuccessful decade for British industry. Efficiency slipped further, management control disappeared as inflation went into orbit and industrial relations, in some sectors at least, descended into farce. As in the 1950s and 1960s, though, companies got away with their uninspiring performance. Despite the oil shock of 1973–4, world trade continued to boom through the decade, providing a cushion into which inefficiency could discreetly sink. It was not until the end of the 1970s that anyone realised that the cushion was about to be whipped unceremoniously away – and by then, it was too late for many companies to avoid disaster.

Heath entered office determined to remove one cushion at least: the one that government provided for industry. He dissolved the IRC and in November 1970 demonstrated his new toughness by refusing to bail out the Mersey Docks and Harbour Board. He also set about controlling the power of the unions: the 1971 Industrial Relations Act introduced secret ballots and cooling-off periods before strikes; it told unions they had to register to keep their privileges. But when push came to shove, the prime minister found it impossible to keep up his tough act.

In 1971, Rolls-Royce went bankrupt. The motor car group was floated off as a separate company, but the giant aerospace division was put into the government's care. If Slater was the ultimate bean counter, Rolls-Royce was run by dedicated spanner brains. The RB211 engine, which it had been developing, was a brilliant piece of machinery whose costs had been allowed to spiral wildly out of control. Spurred on by two world wars, Britain's aviation industry had never lacked technical skills. But it suffered, in common with so much else, from a lack of balanced management.

In 1972 full-blown interventionism was brought back with the Industry Act, which allowed the industry department to pump money into any projects it wanted. Heath did not have the stomach to give the market its full and ruthless head. At the same time, the chancellor, Anthony Barber, made a dash for growth, pouring money into the economy in the hope that it would stimulate self-sustaining industrial investment. Investors pumped their money into property instead. Asset strippers and speculators like Jim Slater and John Bentley flourished. The young trainee John Ashcroft was fascinated by them. "I had a spaniel called Slater and a retriever called Bentley," he says. "Every time these people did a deal I used to analyse it to see how they made it work."

But it was the battle with the unions that led to Heath's undoing. A miners' strike in early 1972 led to blackouts and by the autumn, when the world was transfixed by the Munich Olympics massacre, the unions piled on the pressure – power workers, dustmen and miners all asked for pay increases that were double or treble the inflation rate. The government went on the defensive, and imposed a wage and price freeze: in early 1973, the Counter-Inflation Bill was published, creating the thoroughly interventionist Price Commission and Pay Board. Heath's attempts at free marketism were receiving the last rites.

Throughout that year, the unions tried to smash the freeze with a series of strikes. In September, the TUC expelled twenty unions that had registered under the Industrial Relations Act. The miners started an overtime ban that led, in December, to a three-day week for all industry: the economic crisis had been severely aggravated by the near doubling in Arab oil prices that followed the Yom Kippur war in

October. As the stock market collapsed and the miners declared an all out strike, Heath declared a state of emergency. In February 1974, he called an election, and asked the electorate "who rules the country". He was told he did not, and Harold Wilson came back to power. Immediately, the miners were given a 35 per cent pay rise, and inflation headed upwards.

In the year after the election, average earnings rose by 28 per cent and inflation by 25 per cent. The price of steel rose by 45 per cent, and the price of petrol doubled. The economy was out of control. In the summer of 1976, as the temperature in London hung in the nineties, the government took begging bowl in hand and went to Washington. With banana republic status officially conferred, Britain was baled out by the International Monetary Fund, on condition that it behaved itself. In IMF terms, that meant public spending must be controlled: monetarism was making its bow.

Whatever the government did to control inflation seemed to make things worse. The 1974 Social Contract between the unions and government was supposed to restrain pay increases, but proved worthless, and the government did not help itself with a series of acts in 1975 and 1976 that tilted the balance of power further in favour of the unions. The most controversial, launched just before Wilson handed the prime ministership over to James Callaghan in March 1976, reinforced the role of the union "closed shop". Even the Health and Safety at Work Act was seen by employers as yet another imposition.

The government also attacked inflation by trying to forbid it. In 1974, the Department of Prices and Consumer Protection was set up and the Price Commission, established by the previous government, was instructed to veto inflationary rises.

The trouble was, it rarely did. Big companies had no difficulty in pushing price rises through the Commission. Then, as Sir Denys Henderson, now chairman of ICI, says, "you went to your customer and said, sorry old boy, the Price Commission says we should do this". Financial control in most companies was weak anyway: inflation shrouded already hazy accounts in a dense mist, and removed managers even further from business reality.

Although unemployment was starting to rise – it reached one million in 1975 – the government would not contemplate using it as a weapon against the unions; full employment was, after all, the ultimate aim of its policy. Instead, companies were encouraged to keep labour on. The Temporary Employment Subsidy, instituted in 1975, subsidised companies in development areas that were prepared to postpone redundancies. Proposed plant closures or major redundancies led to endless meetings with civil servants and ministers. "You had to argue your case," says Henderson. "If you could produce the argument, you were allowed to do what was sensible. It just took a lot of time."

After the Social Contract came a compulsory pay policy, which companies were expected to enforce. The government could punish spendthrift employers by telling the Price Commission to block their price rises. But the pressures were mounting all the time, for the policy relied on a flat rate limit (first £6 a week, then £4) which reduced skilled workers' differentials and built up a log-jam of resentment. Even if managers sympathised with the workers' demands – as they frequently did – they were not allowed to give in to them. In the summer of 1977, as the country was busy celebrating the Queen's Silver Jubilee, Lucas lost more than £300 million in a ten-week toolmakers' strike. "The government put pressure on us," says current chairman Tony Gill. "Callaghan was trying to force in something quite unacceptable."

Until the log-jam broke in 1979 with a rash of public sector disputes, there was not an overall escalation in strikes, although some, such as the one that blacked out the state opening of Parliament in 1977, were embarrassingly public. The average number of days lost in 1975 to 1978 was not much higher than during the 1960s. But that was mainly because managers became experts in instant arbitration – at the expense of their real jobs. The motor industry, where manufacturers were reliant on several hundred supplier companies (many of whom would have their own labour problems), was particularly hard hit. "In a typical day, a manager would spend 60 to 70 per cent of his time either dealing with supplier problems or endeavouring to keep the workforce within the four walls," says John Gilchrist, who then worked for Leyland Vehicles.

Says Gill: "We had managers having far too much of their time taken by battling out industrial relations problems, then trying to recover output. Some of them were punch drunk – they weren't thinking about the long term, they weren't even really thinking about today because of the problems created yesterday."

Gill was running CAV, Lucas's fuel injection division, when it took over a company based in Finchley, North London. "The people there were paid more than anyone else in Lucas," he says. "It was run by a man who couldn't stand the cost of saying no.

"We went round to all our customers and said we are going to take this lot on, and that they had better source from somewhere else. It was on and off for eighteen months, disruptions, wildcat strikes, we would take a seven-week strike then be back at work again. I used to have the chairman screaming down my ear, and customers screaming down my ear, although some of the more subtle ones took me out to dinner to persuade me to give more than the other fellow."

In engineering companies, shop stewards had taken on the role of first line managers. Management accepted this, and even found it useful – they did not even attempt to talk directly to the workforce. But it did mean that life became a constant round of petty struggles. Every time the line could not start because of absenteeism, for example, the shop stewards could insist on lengthy negotiations before workers were drafted in from elsewhere. Demarcation lines were defended, flexibility was resisted.

The political element, while played up by the press, was an irritant. "We had a group of Poles and Pakistanis at a plant in Finchley," says Gill of Lucas. "They had all sorts of deprivation to fuel their anti-management feelings. The political motivators would lock on to these people, feed them something they wanted to hear and get industrial action." It did not take many committed politicised people to cause havoc. When the engine test facility at CAV was computerised, union resistance was organised by one man. "He created so much trouble we assigned a man full time to counter his disruption," says Gill. "But eventually we had to close the place and move it somewhere else."

And then there were the union leaders who combined theatricality with militancy: the approach later brought to a fine art by Arthur Scargill. Derek Robinson – Red Robbo – was the best known. He was senior convenor at British Leyland's Longbridge plant and the company, not entirely plausibly, accused him of being singlehandedly responsible for thirty months of ferment in 1978 and 1979, involving an impressive 523 disputes and the loss of 62,000 cars.

Gill believes that the system tended to bring such people to the fore. "It's always been a thankless task to lead anybody on an unpaid basis. If the only rewards are an ego trip, you tend to get the wrong people in the job."

Not all managers, of course, spent their time running around trying to avoid strikes. Some industries, such as chemicals and textiles (where wholesale restructuring was taking place with union cooperation) were remarkably peaceful. Carefully hammered out negotiating agreements brought sudden peace to the shipyards from 1976. And even in car companies, only a section of the management was coping with the shop floor. For the others, inflation had quite a different meaning: it led to an easy life, where money took on an Alice in Wonderland aspect, and profits could be made by doing absolutely nothing. Simply by buying stock and sitting on it, a 15 per cent profit would register in the books. No one was fooled by this, but it suited many to pretend they were.

Direct interventionism flourished under the industry minister, Tony Benn. His pet projects were the Meriden and Kirkby Cooperatives, worked-controlled factories that never took off; more conventionally, he nationalised the shipbuilding and aerospace industries in 1977. Capital investment was pumped into the increasingly unprofitable merchant shipyards – but no suggestion was made of reducing capacity.

Before that, in 1975, the IRC was back: this time it was called the National Enterprise Board. Originally intended to take control of key companies to guide the economy (on the model of the Italian IRI), it did back a handful of high tech start-ups: in 1978, it started up Inmos to make mass production microchips with unique British-developed technology. Inevitably though, its main role was as a hospital for lame ducks.

With these it did try to be commercial. Sir Leslie Murphy, the NEB's first chairman, said the board should be able to nurse companies through dizzy spells – a role undertaken by commercial banks in Germany but not by those in Britain. In one case, Ferranti, it succeeded brilliantly. It also put up most of the money for Clive Sinclair's pocket television after his Sinclair Radionics had got into trouble. As now, Sinclair suffered from the opposite problem to that of most British managers – he couldn't stop launching new products.

One great name faded away in the NEB's care. The board paid 2.5p a share for the machine tool company Alfred Herbert in 1975: in the mid-1960s, the share price had been £18. Herbert himself had died in 1957 at the age of 91; he had been a typical engineer paternalist and had failed to lay the basis of a succession. Those who followed him did not bring in new products, modernise factories or set up a proper control structure. New companies were bought, and were run directly from an increasingly arthritic centre. Soon Alfred Herbert's products were being undercut by better foreign machine tools, and inevitably, the company sank into the quagmire. The NEB tried to keep it going as a whole, and failed; it finally went into receivership in 1983.

What was left of the British machine tool industry seemed scarcely healthier. Starved of investment and new products, it retreated down market. A study in 1979 showed that the average British machine tool sold to developing countries was worth $1,628, against $4,951 for its German rival.

Then there was British Leyland, the ghost that would keep on appearing at state banquets. Since its formation in 1968, Donald Stokes had struggled vainly to control his monster. It had produced a series of new cars each, generally speaking, less successful than its predecessor. It was reckoned to be overmanned by at least 40,000 employees, it produced nineteen models against Ford's five, and achieved less than half the productivity. Even in the late 1970s, British Leyland's accounting systems could not tell whether a vehicle was making money or not. Its industrial relations were farcical. And, when the first oil crisis hit car sales, BLMC went into a spin; inevitably, it became one of the NEB's first charges.

Don Ryder, chairman of Reed International and industrial adviser to the Labour government, was asked to produce a report on the future of British Leyland, which he did in March 1975. He recommended that the government should invest £2.8 billion over seven years. BLMC should be able to keep 33 per cent of the market, said Ryder. His assumptions, it emerged, were hopelessly optimistic and did not tackle the problem of appalling productivity. But by the time that became clear, his plans were roaring ahead.

Ryder's plan was expansionary: the workforce in 1977 was 4,000 higher than in 1975; the company had fifty-five manufacturing plants and new ones were being built. But a toolmakers' strike in February 1977 forced it to the edge of bankruptcy again. By then, Michael Edwardes, chairman of Chloride and a member of the NEB, had taken over. A tough South African, an outsider, he was able to cut through the cosiness and soon slammed British Leyland into reverse. He knew he had to cut, but also identified why the merger had failed so badly: individual units had been stripped of their identity and autonomy; they had lost pride and motivation. The Browns Lane Plant Large/Specialist Operations was given its name back – Jaguar Cars. Edwardes introduced psychological tests for managers, and forced many of them to leave. In 1978, he announced the first major redundancy programme – 12,000 jobs would have to go.

But the chaos continued. Technical problems meant that the new Rover SD1, Car of the Year in 1977 and built in a new factory, was wildly unreliable, while the Series 3 Jaguar was available in only three colours in 1978. Industrial relations got worse and worse. According to Ray Horrocks, the new managing director of Austin Morris, "The difficulties we had with the labour force meant that you had to invest 150 per cent to get 100. The saddest part of it was the general distrust of management."[2] In 1978, Speke number two factory in Liverpool, where the new and unsuccessful TR7 sports car was being built, was closed – the first shutdown in the group since 1948.

The misery continued into the consumer sector. In 1970, British producers sold 182,000 automatic washing machines; importers accounted for about 70,000. By 1978, the market was a million, and half of them were foreign. As with cars, the stop-go policy on credit controls was largely to blame. Thorn had been wrongfooted by Barber's removal of controls on television sets in 1972, and had scrabbled desperately to hold its market share as sales of colour sets doubled. The management lost control of quality and reliability, and Far Eastern companies took 20 per cent of the market. It was becoming clear that the next new appliance would be the microwave, but British manufacturers steered well clear of it, and the Japanese were given a clear run.

A study published by the European Commission in 1979 put the problem in focus. Germany and Japan had preserved or built up a lead by switching steadily to imports for goods that had a low content of skilled labour, and exporting those with a high skill content. If British industry was moving at all, it was in the opposite direction, towards the low wage, low productivity mindset of the third world.

Pressure was building throughout the decade from the Japanese. After the war, Japanese companies, encouraged by the government, had concentrated on labour-intensive industries such as textiles and shipbuilding; but, as wages rose, they had switched to a new philosophy, of using focused or laser beam attacks to carve out market share. One of the earliest victims of this approach was the British motorcycle industry.

In the 1950s, 70 per cent of all the motorbikes in the world were British – the great names, Triumph, Norton, Villiers, BSA, dominated world markets. During the 1960s, the Japanese moved in, first at the bottom end of the market, then gradually higher up. As the Japanese brought in a new machine, the British would look at it, decide they could not compete, and retreat from that segment. By the end of the 1960s, they were concentrating on big bikes: in 1969, of the eight machines between 450 and 750cc available in the USA, four were British – they took a 49 per cent share of the market. But already the Japanese were moving in on this last refuge; by 1973, only two of the ten 450–750 cc bikes in America were British: their market share was 9 per cent. Even as the Americans embarked on a love affair with the big bike, the British, who had dominated the market, failed completely to capitalise on it. While their sales stayed at about 30,000 units a year, those of the Japanese shot up from 27,000 to 218,000. In the first five years of the 1970s, as the standard British reaction to trouble –

merge and merge again – led eventually to one incoherent group called Norton Villiers Triumph, UK bike makers lost about £20 million between them. NVT itself had not long to live.

According to a report produced for the government by the Boston Consulting Group in 1975,[3] the fundamental strategic error was to retreat when the Japanese first moved in with their small bikes in the early 1960s. The only way the British could have competed was by producing in similar volumes to the Japanese to spread higher costs of research, development and sales among more bikes. They should also have followed the Japanese practice of keeping the range of models as small as possible, to generate maximum volumes of each bike. But to achieve those volumes, profitability would have had to suffer in the short term – and that was something British managers (and, more likely than not, their bankers) were not prepared to contemplate. Their models were priced to make a particular profit according to their accounting systems, but, said the report, "the accounting system will be based on existing methods of production and channels of distribution, and not on cost levels that could be achieved under new systems and with different volumes". In other words, the British philosophy was static: it considered costs as given, rather than as things to be reduced. As the short term took precedence over the long term, the inertia in the British industrial system was reinforced.

As a result of the difference in volume, Honda could afford to maintain a 1,400 man subsidiary to manufacture special, and secret, machine tools, while British companies were left with ageing general purpose equipment. At the Wolverhampton engine factory, BCG reckoned that 60 per cent of the machines were more than twenty years old, while investment per man was running at a third of the Japanese level. Similarly, Honda and Suzuki each employed between 800 and 1,300 workers in research and development, while the entire British motorcycle industry had about 100 people. As British machines became increasingly out-of-date, their sales appeal was limited to the real enthusiasts who wallowed in rugged unsophistication. Most buyers, unfortunately, preferred smooth engines and electric starters.

As the quality of British goods fell, so did the morale of British managers. They became even more detached from the business of making money: many scarcely tried any more, and concentrated instead on enjoying their long lunches. As the government restricted wage increases and charged 83 per cent tax on top salaries, fringe benefits – including lavish lunches, expensive cars, boxes at the opera – became more and more widespread. Managers were doing less managing than ever and, an American businessman commented, "the British are the only managers I know who make a habit of boasting to you about how *little* work they do. Lots of British managers just want to get back to their place in the Cotswolds, to their daughters in the pony club, and all the rest."[4]

Big company practices were mimicked throughout industry, with small companies showing they could rival their giant cousins in any waste race. Owner/managers of small companies found they could survive quite nicely on the perks their firms could give them. The company might be on its last legs, but it could always manage a Jaguar, a mistress, a fur coat and a couple of expensive holidays a year for the boss. Everything would disappear into expenses.

There were, of course, good and professional managers coping as best they could. There were companies that flourished in the 1970s: Hanson and BTR were building their empires, albeit mainly overseas; GEC and Racal were managing well, and starting to eye up the weaker electronics companies, Plessey, Decca and EMI. Other managers were learning the lessons of professionalism. Gareth Davies was chief accountant at the West Midlands metal-bashing conglomerate, Glynwed. In 1974, he installed a cash regime, insisting that all units had to give head office 25 per cent of their assets in cash every year: as a result, it had built up its own cushion when the downturn came. Davies was a rare man in 1970s Britain.

During the 1970s, the gap in efficiency between British industry and its competition widened into a chasm. But again the health of the world economy, which had grown with remarkable strength since the Second World War, came to the rescue. The first oil shock caused a hiccup, but as the oil producers' new wealth was recycled into vast capital projects, Western contractors and suppliers were given a bonanza. From 1976 to 1978, world trade expanded rapidly, and British companies won a sufficient share to keep healthy. The average rate of return for non-oil companies crashed from 9 per cent in 1973 to 4 per cent in 1975 but by 1978 had recovered to 7 per cent. By that year too, the Callaghan government's pay restraint policy, combined with IMF-imposed restrictions on public spending, had brought inflation down to below 10 per cent for the first time in five years. The British economic edifice once again looked respectable.

But it was a house built on the creakiest of industrial foundations, and forces were massing to undermine them. The most dangerous of all should have been an ally: it was called North Sea oil. In 1970, BP made the first big find in the British sector of the North Sea. In November 1975, oil started coming ashore and by 1978, more than half of Britain's domestic needs were covered. There was a mini oil-based boom as North Sea platforms were set to work from 1976 to 1979: the oil supply industry accounted for a quarter of all British industrial investment, and fed work to a number of boom towns on the east coast. (There could have been more: BP complained that it could not get many of its needs from British suppliers.) But there was much discussion over the true effects of North Sea oil. Would it enable Britain to rebuild her industrial base, or would it somehow speed her decline?

A sign-off to the Boston Consulting Group's report on motorcycles got it right in 1975. "There is the risk," it said, "not too far fetched, that North Sea oil earnings will be flowing within four years to the extent that the pound's value is boosted. This will make it that much harder for British manufacturing industry to compete, for the pound's value will not be a true reflection of the country's manufacturing effectiveness." Although UK oil production increased from 1.6 million tonnes in 1975 to 80 million in 1979, it was not until the second oil shock, triggered by the fall of the Shah of Iran in January 1979, that the BCG's prediction proved absolutely and devastatingly accurate. Foreign exchange dealers started to label sterling as a "petrocurrency", and pushed its value relentlessly upward.

One of the reasons that British industry had survived in the past was because it had low wages compared with competitors: that was why companies could tolerate overmanning, and could underinvest in labour-saving equipment. In economic theory, a currency should decline if inflation is ahead of its rivals: and sterling had indeed been spiralling down as competitiveness waned (in devaluation steps before fixed exchange rates were abolished in 1973; in freefall thereafter). Now, the spiral was broken. With wage inflation powering UK costs up, and sterling appreciating at the same time, British industry was heading rapidly for the edge of a cliff.

The second oil shock had another, broader, effect: it put a brake on the world economy. After the recovery from the first shock, most managers assumed that the unbroken growth since the Second World War was back on track. But it wasn't. Supply was slipping dangerously ahead of demand – not least because huge new chemical and steel plants, funded by the first round of oil price rises, were now coming on stream; third world countries that had no oil were heading towards bankruptcy, encouraged by Western banks desperate to lend them money; and the economic powerhouse of the United States was starting to falter as oil costs slammed home. It would not take much to tip the whole system into recession.

But the first cracks to appear in the British economy were forced open by domestic pressures, and particularly by the Callaghan government's attempts to control pay rises. In the summer of 1978, a 5 per cent limit on pay rises was imposed. As winter approached, it became clear that this would push the increasingly strained incomes policy too far. At the Labour Party conference in October, the unions condemned the limit and in November, after a nine-week strike, the Ford workers broke it with a 17 per cent rise. That month's bakers' strike was far more worrying for the average Briton, but among managers there was no doubt which dispute was the more significant. The Ford dispute marked the start of the Winter of Discontent.

On January 5, 1979, the lorry drivers began a strike in support of a 25 per cent pay claim: it was marked by violence and secondary picketing, and caused shortages of heating oil and fresh food. Then came a string of public sector disputes, which created the sordid havoc that gave the press a photographic field day: ambulancemen, sewerage workers, municipal employees, school janitors and dustmen all came out to protest against the 5 per cent limit. The unions were deeply unpopular, even among their members. In the election – triggered not by the unrest but by the government's refusal to pass Scottish devolution legislation – more skilled workers voted Tory than Labour, for the first time since the Second World War.

# 3

# *Tory philosophy and the first year*

Tory economic policy came largely from Sir Keith Joseph. He was, as one industrialist said, "a nice man, but a bit twitchy to say the least". He came from a wealthy business family – it owned Bovis – but, as a Fellow of All Souls, he had little enough in common with the average businessman. His extraordinary mannerisms and utterly cerebral approach had got him into trouble more than once – most spectacularly when he announced in 1974 that lower-class women were having too many children. He must have had more rotten fruit thrown at him than any other politician. But his economic ideas, refined from 1974 at the Centre for Policy Studies, were straightforward enough.

Joseph believed that Britain's difficulties came from the supply side. To traditional Keynesians, the rising unemployment of the 1970s could only mean that there was a lack of demand: government should therefore increase its spending, and industry would respond. But, at the end of that decade, there was no shortage of demand: indeed, ever higher real wages meant that consumers were having a high old time – as long as they kept their jobs. Yet industry was not responding. In 1978 and 1979, real incomes grew by more than 10 per cent, while output went up by only 4.5 per cent. Joseph believed that the problem, and the solution, lay with industry, not government.

No amount of government legislation or pacts would control wage inflation. Only economic pressures could do that and economic pressure worked not on Whitehall, but on the boardrooms of British business. If enough pressure was put on businesses, they would have – for their own survival – to hold down the rise in "unit labour costs", wage rises less any increase in productivity. Unlike traditional Tories, he believed government had almost no positive role to play. Its job was to restrict the amount of money in circulation, and to spend less: that would help to control inflation in itself, and would also supply the necessary pressure on business by pushing up sterling. The government should also do everything it could to destroy the power of monopolies, whether they were wielded by nationalised industries or by trade unions. If it was to be involved directly in industry, it should be only to encourage the birthrate of new businesses and also, possibly, new technology.

With inflation under control, Joseph believed a new dynamism should return to the economy. But he was well aware that the whole policy depended on managers suddenly finding new strength, and new entrepreneurs emerging in droves from the woodwork. He was not terribly optimistic that that would happen. He talked of "punch drunk and patchy" management as a root cause of Britain's decline and noted that "as a nation we seem to have lost much of the entrepreneurial edge which once made us the wealthiest nation in the world".

At the age of sixty-one, he was put in charge of the Department of Industry. He hurled himself into his job with gusto, taking himself off on lightning visits to Glasgow, the North-East and the North-West. He had breakfast with Cammell Laird's management and visited a nightshift at a Northern Engineering Industries factory in Newcastle. His manner was described by John Elliott of the *Financial Times*. "His intense inquiring style often pulls people out of their prepared speeches and into some acceptance of his ideas. He draws up an instant agenda and launches into staccato, but elegantly polite questioning as ideas strike him." He also displayed his well-known tactlessness, telling pickets facing a factory closure in Scotland that 90,000 people changed jobs every week, and offering to send them details in the post. His own conclusion was that he "found a fairly widespread assumption that jobs come from government – yet in the same areas I've met impressive examples of the successful entrepreneur".

In his office in Westminster, Joseph found his fundamental free marketism being undermined from the start. He arrived determined to slash regional aid, which he thought did no more than shuffle investment between areas. "There should be a monument to the unknown unemployed, the men who have lost their jobs because of state aid elsewhere," he liked to say. And he wanted to reduce the role of the National Enterprise Board to that of nursemaid for the four real basket cases: British Leyland, Alfred Herbert, Cambridge Instruments and Rolls-Royce.

## Tory Philosophy and The First Year

One of his first moves was to give his civil servants a reading list of books on the damage of state intervention. Two and a half months later, they got their own back with a dossier full of figures on state aid given to industry in Britain's major competitors, including the USA and West Germany. His assistants were surprised at the speed with which he started backtracking. "He sees the counter-argument almost too quickly," one commented.

As a result, his removal of "market distortions" was less than wholehearted. He cut regional aid by more than a third, but left its structure intact. He forced the NEB to sell off its profitable holdings – in ICL, Ferranti, Brown Boveri Kent and Fairey Holdings – but left it with £20 million a year to invest in small and regional companies. It would also be allowed to keep its four new high tech ventures, including Inmos. "Institutions may not be ready to take the risks they should," he conceded. By the end of the year, he had let the NEB spend another £220 million on its lame ducks, and Inmos had been given the go-ahead to build a £24 million British production plant (so far, it was operating only in Colorado). He also said that the government would continue to subsidise British Shipbuilders' losses, even though not a single merchant shipyard was making money.

In only one industry did Joseph take a hard line. He told British Steel that the government would not fund its losses after March 1980: they were currently running at £1 million a day. There was no secret about what needed to happen to steel: a plan to rationalise BSC into five coastal sites had been embarked upon in 1973. It had made desperately slow progress (steel plants sat on more than fifty Labour seats), and in any case was now out of date. It was based on projected demand of 36 million tonnes; by the end of the 1970s, actual demand was 15 million tonnes. The minister told the BSC to go ahead with its plan to close the main works at Shotton and Corby, throwing 12,000 people out of work.

Away from Joseph's direct orbit, the new government was moving fast. The chancellor, Sir Geoffrey Howe, whacked up interest rates in an attempt to control the money supply and suppress inflation: in June they jumped from 12 to 14 per cent, and in November they leapt again to 17 per cent. Planned public spending was pruned vigorously. The Price Commission was abolished, so were dividend controls. Top tax rates were brought down from 83 to 60 per cent, with basic rates dropping from 33 to 30 per cent. To compensate, indirect taxes were increased: VAT went up from 8 or 12.5 to 15 per cent. In November, exchange controls were abandoned – in the long run, the most significant move of the year for industry. And, while wholesale denationalisation was not part of the Conservatives' agenda, ministers announced that part of the state's holding in British Aerospace would be sold off.

The VAT rise in June did little to vanquish inflation, which was also stoked up by rising oil prices. As the Iranian revolution rolled on through the year (the Teheran hostages were taken in November), the contraction of supplies allowed OPEC countries to ratchet up their prices. At the end of 1978, crude oil cost about $12 a barrel; 12 months later, it cost $26. As the unions plunged into the first pay round of Thatcherism, they saw no reason to be moderate, especially as the government had announced sweeping proposals to restrict their power. In August the engineering union, AUEW, called a series of one-day strikes for higher pay and a shorter week. The next month, they were upgraded to two-day stoppages, and engineering firms started to close plants and lay off workers. In September, more working days were lost than in any month since the 1926 General Strike. As the autumn wage round wore on, the pressure on wages – and thus inflation – built up. In November, the miners demanded a 65 per cent rise. During the year, average earnings growth went from 13.5 per cent to 18 per cent; inflation rose from 8.4 per cent to 17.2 per cent.

A call at the TUC Congress to start mass demonstrations and resistance was only just defeated. Tom Jackson, the moderate leader of the Post Office workers union, told Congress that "we cannot and will not restrain wage demands when the government has abandoned all attempts to control the rocketing level of prices, withdrawn from consensus policies and deliberately fostered unemployment".[1] Mrs Thatcher countered at the Tory Party conference. "Millions of workers," she said, "go in fear of union power."[2]

As the decade grew to a close, it was the twin demons of British Steel and British Leyland – the 1970s equivalent of the Balkans Crises – that were occupying the vast mind of the secretary of state for industry.

The rising pound through the year had reduced BL's prospects from flickering to dim. Although Edwardes had signed an agreement in May to build a rebadged Honda model, and was gearing up to launch the Metro in 1980, BL was losing money on exports, including £900 on every MGB going to the USA. At the same time, home market share had dropped from 31 per cent in 1975 to around 20 per cent. In September, Edwardes announced a recovery plan that would involve the shedding of 25,000 jobs and the closure or partial closure of 13 plants, including MG at Abingdon. Mrs Thatcher had made it clear that she felt no especial need to keep British Leyland going, so Edwardes decided that there was no point in being subtle. He gave notice that he was taking the unions on by balloting the workforce directly on his proposals: 80 per cent supported him and, when Derek Robinson stood up and said they should in any case be resisted, he was sacked. Edwardes had passed the first toughness test: Sir Keith Joseph swallowed hard, and handed over £300 million.

BSC's troubles were closely linked with those of BL – the car company took 13 per cent of its output of sheet steel. In October, it announced it wanted to start running down the Shotton plant in December – a year earlier than planned. Ten days before Christmas, the most sweeping rationalisation plan ever was proposed: within eight months, 52,000 jobs would be lost. Every plant would have to cut back and the works at Consett, in County Durham, and Hallside, in Scotland, would close. On December 13, the Opposition introduced a motion that Sir Keith's salary should be reduced by £10,000 as a form of censure. It was defeated.

In the Commons, Nicholas Edwards, the Welsh Secretary, said he was "shocked and dismayed by the scale of the industrial and social situation with which we are faced". He also demonstrated an optimism which, with hindsight, seems dazzlingly sunny, by predicting that BSC would be in profit in the financial year 1980–1. Thanks to the next year's events, he was a billion pounds out.

In London's Oxford Street, there were sure signs that recession was on its way. Christmas shopping, which usually started around the end of October, was still struggling into life in early December. After the consumer boom of the last two years – caused simply by the rapid rise in wages – the government had pushed interest rates up sharply. With unemployment rising, the British people started hoarding.

It all proved too much for a toy company that symbolised old-fashioned British quality – Meccano. Alec Issigonis had modelled the automatic transmission of the Mini in Meccano; Harold Wilson's first office had been as president of his local Meccano Club. But, said its owner Airfix, the problems of the Liverpool factory were familiar ones: failure to update the range, explosion of costs, labour problems, competition from the Far East. Everything, in short, for which the 1970s stood. Management closed the factory at forty minutes' notice at the beginning of December, accusing the workforce of sabotage, drunkenness and anarchy. The unions countered with accusations of management inefficiency, financial irregularities and blackmail. As the workers occupied the factory, His Holiness Maharishi Mahesh Yogi appeared, and said he wanted to rescue it if the workers took up the meditative path to recovery. It was a suitably bizarre end to a bizarre decade.

# Part Two

## *Depression*

# 4

# *Towards the cliff edge*

On January 2, 1980, Bill Sirs, leader of the Iron and Steel Trades Confederation, called his members out on strike. For the next three months, not a tonne of steel was produced by the British Steel Corporation. It was the first major strike in the industry since 1926.

The immediate issue was not the massive rationalisation plan that had just been announced: it was a dispute about pay and working practices. BSC had offered 2 per cent, soon upped to 6, while the ISTC wanted at least the same as inflation, around 17 per cent. The corporation said it would offer more, but only if local productivity deals could be set up, and petty demarcation abolished. As in most big companies, silly practices had been allowed to survive after technology had made them inappropriate: a fully trained electrician was needed to change a lightbulb; only craftsmen could change steelcutting blades, even though it was a simple job with a spanner. They were all old-fashioned union privileges, and Sirs was an old-fashioned union man.

Not that BSC's bosses were highly regarded. A disaffected manager claimed in the press that they were desperate to cut capacity that they had foolishly built up on the basis of wildly optimistic forecasts. The 2 per cent pay offer was deliberately cynical, and was designed to precipitate a strike that would weaken the workers and the industry. The top management, said the employee, lacked leadership, suffered from bureaucratic malaise and preferred confrontation to negotiation.

Almost immediately, mass picketing blocked the steel works. "They failed, we pay," said a picket's poster. Docks were picketed; so were warehouses; private steelmakers were blockaded. A consignment from the Consett works was trapped at Wolverhampton Rail Freight Depot in a highly organised secondary picketing operation. Stockholders started to ration their supplies. Within a week, the first casualties, 100 men at York Trailers, had been sent home.

Soon the strike was mushrooming into a political dispute, as Joseph made it clear that he was not prepared to give any extra money to fund a pay settlement, or to write off any of BSC's debts. "The steelworkers are not being sensible in insisting on higher pay regardless of whether or not productivity has improved," he told the House of Commons, metaphorically waving his textbook. They did not want to be sensible, they just wanted to win. They jeered him when he visited the Midlands, and stepped up their secondary picketing. Sometimes it became violent, especially at the private steelworks which had been dragged into the dispute.

The unions had no doubt that this was the first serious challenge to their musclepower since Edward Heath had taken on the miners in 1974: they wanted to prove once again that they could not be suppressed. The ISTC told its unions to halt all steel movements. If BSC decided to close South Welsh operations, it would be "engaging the wrath of the TUC", Sirs warned. He implied that a general strike was a distinct possibility.

As the strike deepened, users were being forced on to short time. In February, Metal Box began laying off 2,500 workers. Joseph tried to go on a factory visit in Wales, and was pelted with rotten eggs and fruit. A tomato hit him on the chest. His attempts at dialogue were less fruitful. A Gwent County Councillor, also a steelman, shouted at him: "You are mad. We hate you."

By March, the strike was becoming messy, with private companies back at work and imports coming in through East coast ports. The ISTC issued a list of forty-two firms that had been forced to close or lay off men, but car firms were managing well enough – even British Leyland was buying foreign steel. BSC's offer had crept up to 10 per cent plus 4 per cent through productivity schemes, and in the middle of the month both sides agreed to set up a committee under Lord Lever to arbitrate. It recommended a rise of 11 per cent in return for an agreement on working practices, flexibility between jobs and local productivity deals that could add another 4.5 per cent.

It was a damp squib ending to a bitter strike. The workers got most of the money they wanted; the corporation got most of the changes in working practices. The unions indicated that conflict was not over yet. "If we don't get any of these closure decisions lifted there is going to be big trouble," Sirs threatened.

The main beneficiaries were the foreign steelmakers. They had all been suffering from overcapacity and swooped like starving vultures as BSC languished. Desperate British customers had to accept long-term contracts with them: the corporation thought it had lost 10 per cent of its market permanently. It was, Ian MacGregor wrote, "a strike that must go down as one of the great lemming runs of all time."[1] Not that the management emerged with much kudos: why, people asked, had it made such a miserly initial offer when its later offers would probably have been accepted without a strike?

With the benefit of hindsight, the strike was clearly the turning point for BSC. At the time, the dispute was seen as a political smoke signal from the government: however much it had intervened behind the scenes, it had at least not been seen to be working for compromise. The unions were being sent on a course to learn how to behave, and this was their first lesson. Coincidentally, it was during the strike that the Tories introduced their first bit of union-bashing legislation, the Employment Act announced the previous autumn. It was designed to restrict secondary picketing and closed shops, and to insist on secret ballots before strike action. Designed by the unreconstructed Heathite, or wet, Jim Prior, it was too mild for much of industry, but reflected well enough the views of the average Briton. A survey published in January had shown that 68 per cent of unionists thought that trade unions had too much power, and two thirds backed the government's proposals to limit picketing. But the TUC was not going to be pushed around: an attempt to allow unions to accept funds for ballots was massively defeated by general council.

Just as dramatic were the battles of Michael Edwardes at British Leyland. Here was a macho manager who took on the full machismo of the union movement (or so it appeared – in fact, he was being bolstered by an even more macho board that included the little-known American Ian MacGregor). In February, he won a small but high-profile victory when attempts to reinstate Derek Robinson as convenor at Longbridge failed. Red Robbo was not a popular man among the workers partly, ironically, because he had been seen as soft during a stoppage the year before. He was also quite unsubtle. "Don't try to sack me because you will see what will happen," he told managers. Had Edwardes not sailed quite so close to the legal wind when he fired him, and had he not trodden repeatedly on the AUEW's corns on a number of other issues, it would hardly have kicked up a fuss.

But Red Robbo was only one of Edwardes' problems. Not only had British Leyland's costs relative to the Germans risen by 20 per cent in the last year, its ageing model range was losing out at home: January market share was 15 per cent, the lowest ever. The company responded by laying off 30,000 workers. Photographs showed wedge-shaped Austin Princesses lined up in fields around the Cowley plant. Leyland launched a massive price-cutting campaign that helped the market share scrabble back to 23 per cent by March, but nobody was that impressed. It was not as though there was even a recession in the industry: overall car sales were up somewhat on the year before. Leyland-watching was as interesting a sport as any, and speculation about the group's liquidation abounded. In HM Treasury, it was more than speculation – detailed studies were being made on the effect. Roughly speaking, 500,000 jobs depended directly or indirectly on the group. Edwardes' recovery plan, agreed by ballot in September, seemed to be heading for the rocks; and Joseph's free marketism was being put to another unpleasant test.

Everybody had been talking about a recession for months now. There were plenty of doom merchants about, who saw how rising interest rates and sterling should hit industry, and were making suitable noises. "It's pretty useless trying to create an efficient steel industry," said Leslie Tolley, chairman of the British Institute of Management during the steel strike, "if the users are not going to be functioning."

The *Sunday Times* published a feature: "Why Britain's industry is falling apart." A household scene was photographed, with the caption, "The only thing made in Britain was the baby". The paper contemplated booming West German industry, noting that German output per head was double Britain's. German industry was moving upmarket, while Britain's was moving down. Britain's plant was hopelessly outdated. We were not producing enough engineers (the newly published Finniston report said there should be a new body to represent them). We were missing the high tech boom. And, the paper pointed out, all these reports argued that there should be more, not less, government involvement in industry. It should be training engineers, it should be subsidising high tech. Import controls were advocated, and not just by the Left. "Temporary controls to provide threatened industry with a breathing space must be regarded as fair," said the CBI.

The first whiff of criticism was emerging from the Cabinet itself. Sir Ian Gilmour, Lord Privy Seal, and founder wet, told a Cambridge audience that "economic liberalism, because of its starkness and failure to create a sense of community, is not a safeguard of political freedom but a threat to it".

Everything seemed to be moving in the wrong direction at the same time. The breakdown in wage restraint of the Winter of Discontent continued through the next winter as unions piled in one claim after another: now that the government was the enemy, there was even less reason to hold back. Public sector workers did particularly well as the Clegg Commission came up with a series of generous payments to keep them in line with private employees: teachers received 18 per cent, nurses 19 per cent. As spring approached, average wage settlements were touching 20 per cent, the highest level since 1975. Inflation was at about the same level, while interest rates were sitting at 17 per cent.

At the beginning of the year, many economists were saying that the pound had to fall: wages and inflation were rising so fast, the balance of payments was going to be in deficit again, and other countries (especially the US) were pushing up their interest rates. They were wrong: the oil-fired pound kept on going up and British industry found itself 40 per cent less competitive than it had been three years before.

What was worse, labour costs had been outstripping price increases by more than 50 per cent for the past five years. In other words, profits were being squeezed: return on capital for non-oil companies had fallen from 7 per cent in 1978 to less than 4 per cent in early 1980. That meant that companies had less padding to cope with these frightening new pressures.

Despite all this, the signs of industrial disaster were remarkably few. Apart from the odd lame duck, it was only companies susceptible to Far Eastern competition that were really feeling the squeeze.

Announcements of major redundancies in the manufacturing sector had been running at between 5,000 and 7,000 a month since the General Election. The closures of Shotton and Corby by BSC were the biggest individual knocks, but by far the most widespread rationalisation was going on in shipbuilding and textiles: 14,600 job losses were announced in shipyards, and 15,400 in textiles. Courtaulds alone shed 5,300 jobs. And after the collapse of Meccano, other toy companies were feeling the pinch. Lesney, makers of Matchbox toys, found it could not compete with Far Eastern makers and made 1,275 people redundant in January. The next month Dunbee-Combex-Marx, of Hornby trains and Scalextric cars, called in the receiver.

Elsewhere, the recession seemed reluctant to arrive. For many firms, the first three months of 1980 were as good as any: ICI reported record sales; the car market was buoyant. A sign of confidence was that stocks of raw material and work in progress remained high. And from December to the beginning of March, the FT30 share index rose by 12 per cent. "We feel that official forecasts of a 2 per cent drop in output may turn out to be too pessimistic," said David Grenier, of stockbroker Scott Geoff Hancock.[2] Even the shipbuilding sector was making optimistic noises. British Shipbuilders may have been losing £100 million a year, but it had been through massive rationalisation, it had several ultra-modern yards, good labour relations and, in contrast to BSC, a coherent strategy. This year, BS needed to win forty-five orders; by spring, it had already landed thirty-three.

If a recession was on its way, newly monetarist governments saw it as their job to hurry it along. Chancellor Howe introduced a savagely deflationary budget in March. For the first time since Keynes had cornered the market in economics, here was a government that reacted to downturn by cutting, not spending. The difference between the Tory budget and what Labour had planned was huge: £10 billion in public spending. Council house budgets were slashed, hurling the construction industry into despair. And nationalised industries were expected to get through £1 billion less than they would have had under Labour. Furthermore, the government had made it plain that it would do nothing to stop redundancies if and when they started. Indeed, it appeared it would almost encourage them.

Shortly after Denys Henderson joined the ICI main board in March, he went to see Sir Keith Joseph to explain that the company was pulling out of polyester fibres, and wanted to close plants at Kilroot in Northern Ireland and Ardeer in Scotland. More than 4,000 jobs would be lost. "I went in on the basis of my experience as Commercial General Manager in the mid-'70s," he says. "So I had all the arguments, and said I have to tell you that we want to do this. He said yeah, OK, that's your decision, bye-bye." The minister did not even ask for the reasons.

On the other side of the Atlantic, Paul Volcker, chairman of the Federal Reserve, was doing his bit to bring recession closer. He was pushing interest rates up and up to try to squeeze inflation out of the economy. As inflation failed to respond, so rates were hiked up again. The higher they were when they started to bite, the sharper would be the effect.

In April, there was a sudden breakthrough at British Leyland. March had been a better month for market share, but the nagging problems of working practices were no nearer to being resolved after five months of talks. The unions saw the company's proposals as an attack on shop stewards' authority, while the company regarded them as essential to improve productivity.

Edwardes had produced a ninety-two-page document setting out his ideas. A key feature was the abolition of "mutuality agreements", which gave shop stewards huge influence in the running of the plant. He wanted that stripped away from them, and power handed back to the managers.

Finally, in April, he imposed the proposals, linking them to a modest 5 to 10 per cent pay package. He used his trick of bypassing the unions by summarising his plans in the "Blue Newspaper", which was sent direct to the workers. Anyone who turned up after Easter, Edwardes said, would be deemed to have accepted the proposals. On the Friday after the holiday, the TGWU called its 47,000 members at BL out on strike, but was ignored by the two largest plants, Cowley and Longbridge. The other main union, the AUEW, told its men to keep working and soon the TGWU leaders agreed to start talking. In the same week that the most modern car plant in Europe was thrown open to view – it would start making Metros in October – British Leyland management seemed to have cured a good part of the arthritis that was causing productivity to seize up. "Thirty years of management concessions were thrown out of the window ... our car factories found themselves with a fighting chance of being competitive," Edwardes wrote.[3]

To managers throughout industry, the steel dispute and the battles of Little Michael, as BL workers called Edwardes, were turning points. For the first time since Heath's efforts of the early 1970s, the balance of power between unions and managers was tipping slightly in their favour. Between them, British Steel and British Leyland symbolised everything that was wrong about industry.

But things were to get much worse before they could start getting better.

# 5
# *Recession*

The recession hit about April, the month of the failed American attempt to rescue the hostages in Iran. The clearest signal was the car market: sales were almost 30 per cent down on a year before. But companies in a range of industries found their order books emptying. Between the second and third quarters, manufacturing output fell by 4.5 per cent – the deepest collapse for 50 years.

The sharpness of the shock was to a large extent the result of the steel strike. With the economy so distorted anyway, it was difficult to detect underlying trends. Private steelmakers, who had been selling at healthy margins after pickets had left their gates, were hit most directly. "We suddenly found ourselves with order books halved," says Constantine Folkes, chairman of Black Country-based steelmaker John Folkes Hefo (now Folkes Group). "We didn't realise the recession was pouring in on us."

In May, while the British people stared fascinated at television pictures of SAS men storming the Iranian embassy, the British economy was having a bad month. Inflation reached 21.9 per cent, pay rises were about the same. The National Institute of Social and Economic Research predicted that total non-oil profits would be less than £4 billion in 1980, against £15 billion in 1979. On May 14, the TUC held a Day of Action, a general strike against government policies. And, to cap it all, there were strong rumours that Sir Keith Joseph, long nicknamed the Mad Monk, had finally gone over the edge. On May 2, he announced to the House of Commons that the government would be paying an American merchant bank, Lazard

Frères, up to £1.8 million to release Ian MacGregor for three years. He was to succeed Sir Charles Villiers as chairman of British Steel, a company that was in the process of making 52,000 people redundant. John Silkin, Labour's Industry spokesman, said that this was "the most staggering statement this House has heard in a long, long time". MP Barry Jones simply asked whether Sir Keith was feeling well.

Although Joseph likened the payment to a football transfer fee – MacGregor's salary would be £48,500 – and pointed out that the bigger the transfer fee, the better the player, the American was not his first choice. Three Britons had been asked, and had turned the job down. Candidates for the job included John Harvey-Jones, deputy chairman of ICI. But MacGregor was not as unlikely a choice as it had at first appeared. A Scots-born metallurgist, he had been sent to the USA in the war, and had stayed to build Amax, a small molybdenum producer, into one of the biggest mining companies in the world. He had been non-executive director of British Leyland since 1977, and was known as a toughie. An outsider like Edwardes, he could cut straight through establishment sensitivities, and had little respect for the unions. "My greatest concern is not how many redundancies we will have, but how many jobs we can salvage," he said. He and Joseph got on well. "I wish we had more Mr MacGregors," the minister told Parliament.

Unemployment did not rocket immediately. Companies moved on to short time and made what savings they could, before handing out notices. In April, only 2,303 redundancies were announced; in May, there were 9,641. At the beginning of June, John Evans of the CBI's West Midlands branch commented that "it is remarkable that so little labour has been shed up to now". By then, a quarter of the 1,200 companies in the West Midlands Engineering Employers' Association were working on short time and, even as Evans spoke, the dam was being breached. In June, 23,556 redundancies – including 3,700 at British Steel's Consett works – were notified. The great manufacturing shake-out had started.

Traditional blackspots were steeped in gloom. In the Northern region, sweeping across from Tyne-Tees to Cumbria, there were 142,706 people unemployed at the beginning of July, and 6,785 registered vacancies. The branch economy was once again proving a curse: the North-East was a region where traditional armaments and heavy engineering industries had been replaced by branch factories of companies with headquarters further south. As the locals knew from bitter experience, it was always these that suffered first when the markets turned down.

But it was the engineering and motor industry that felt the shock hardest. Vauxhall laid off 5,000 people at Luton and Ellesmere Port. The wealthy Midlands autocracy felt the sharp snap of recession for the first time. Lucas announced 3,000 redundancies and two plant closures, Renold and Wilmot Breeden planned another 2,000 between them, while GKN's redundancies took the guts out of Telford, launched as a dynamic new town in 1968.

The Black Country, where British industry had started 200 years before, was the workshop of the engineering industry. Each borough had a speciality: springs were made in West Bromwich, nuts and bolts in Darlaston, locks in Willenhall, tubes in Oldbury, nails in Dudley, chains in Cradley Heath. Although their products were needed throughout industry, car firms had become ever more important customers. For the most part, these firms were run by paternalists – either first-generation engineers or second- or third-generation heirs to the business. Their main fault was an inability to change; but their main strength was loyalty. They would do everything they could to keep their employees in work.

Dart Spring of West Bromwich was typical. It turned out 3.5 million springs every day, ranging from delicate little numbers for hi-fi turntables to massive coils for lorries. Through the winter, pressure from interest rates had been building. Dart's chairman, Arthur Griffin, whose father had been chairman before him, had looked for ways to save cash: he delayed settling bills as long as he could, while demanding payment from creditors as early as possible. The trouble was, everyone else was doing the same, and it was the little guys – like Dart – that were squeezed the hardest.

The fall in output was not caused by a sudden collapse in demand, but by a great rush to cut stocks. During the 1970s, companies had become appallingly lax with their stock control. In the days of high inflation, a good way of making money was to buy stock and sit on it for a while: tax relief made the game more fun. Even in 1979, the level of stocks had risen rapidly. Now, companies started to use up their stocks as fast as they could to save having to reorder. With margins falling and interest rates rising, they would do anything to avoid getting into debt.

The usual bolt hole for companies faced with collapsing orders, an export drive, was firmly blocked by the strength of sterling. By mid-May, Griffin had no choice: 84 of his 650 workers would have to be made redundant. At 5 pm on June 20, they left to join the dole queue, with an average of £350 in their final wage packets.[1]

A few companies had seen the recession coming. Glynwed, also based in the Black Country, had learnt in the 1970s that companies had to be "resized" as the market contracted. In 1978 and 1979, it had started to close factories and, as treasurer Christopher Purser says, "batten down the hatches and run for cash". But even for Purser, "it was a worrying old time, particularly sitting here. One of the ways the recession manifested itself was that you could suddenly get to work ten minutes quicker. I remember driving down Peartree Lane in Dudley; it used to be filthy with smoke everywhere: now it was silent and shut."

The recession also knocked the wind out of consumer industries, particularly those with strong foreign competition. In June, Singer decided to close its 3,000-strong sewing machine factory in Clydebank; a BSR record player deck factory employing 1,600 people would be closed at East Kilbride; Hotpoint announced 477 redundancies at Mexborough; Ward White closed a Northamptonshire shoe factory; a pottery in Stoke cut 120 jobs.

Even companies that were relatively immune from foreign competition were hit. Sweetmaker Cadbury's handed out 700 notices at Bournville, while Rowntree's put 1,550 people on short time. For them, destocking among distributors was the problem, triggered mainly by high interest rates, a desperation to squeeze cash out of any crevice, and a total loss of confidence in the economy.

Coloroll was a healthy firm, having found a nice niche for itself in the Laura Ashley style of wallpaper. It had a dynamic and obsessively analytical young managing director, John Ashcroft, and 1979 had been an excellent year. But in May, Ashcroft says, "the world stopped buying everything. We used to go to Wallpaper Makers' Association meetings and joke about who had got that month's order – it was incredible." Coloroll made 10 per cent of the workforce redundant, then sat tight. Ashcroft reckoned that for him, at least, this was just a very severe bout of destocking.

Few companies were hit harder than Bernard Wardle, the biggest producer of PVC-based leather-look plastic that covered most of British Leyland's car seats. Not only was British Leyland making fewer cars, it was turning more and more to cloth for its seats.

On May 6 Brian Taylor, a burly 47-year-old, took over as chief executive; he believed innocently that this was a stepping stone to greater things. The stepping stone, he soon realised, was more like a Himalayan peak.

That Taylor was not as other managers were had become clear early in his career. Born into a poor Bristol family, he had gone to the local grammar school and university, where he had taken a degree in geography. After two years trudging round the bush of East Africa as a land surveyor, he came back to do his national service in the RAF. He joined Associated Fisheries as a trainee manager in 1959 and within twelve months was picked to run a large fish processing factory in Grimsby; it was losing several hundred thousand pounds a year. Wheeling and dealing early every morning on the fish dock, he picked up basic business principles: that time, money and raw materials had to be juggled to make each day's production run as profitable as possible. "I amused myself doing mathematic models that predicted the price of fish on Grimsby market," he remembers.

After eighteen months, the factory was in profit, and Taylor moved off and up on a career that took him through marketing and into general management. By 1979, he was running Wilkinson Sword's safety and protection division. To each job along the way he had applied ruthless common sense, combined with plain ruthlessness. He was, it was said, a sort of corporate Clint Eastwood.

By the end of the 1970s, he decided he wanted to run his own company, and identified a marine safety company, RFD, as a possible target. He needed a backer, and met by chance a young man called Graham Ferguson Lacey, whose reputation was based on a strange combination of frenetic dealmaking and Christian fundamentalism. Lacey was the last of the 1970s whizzkids in the Slater tradition, and had a knack for squeezing money from the most unlikely deals. In early 1980, he was busy trying to buy a 19 per cent stake in Tiny Rowland's giant Lonrho group. But he had also just clinched a deal to buy a more modest target, Cheshire-based Bernard Wardle. Ferguson Lacey would be delighted to help Taylor buy RFD, he said, but first he would like Taylor to come and sort out Wardle's.

Bernard Wardle was in a mess, far worse than anything Taylor had expected. The end of 1979 figures had not looked too bad; by May, the company was losing £300,000 a month on an annual turnover of £28 million. Its previous management had made belated attempts to diversify away from leather-look fabrics. One small company it had bought, Revotex, had a new sound-deadening product; but when Taylor arrived, the machine needed to make it was lying in bits all over a factory floor in Blackburn. An attempt to move into garden fittings had proved wholly disastrous. The only clear attempts at rationalisation had been a decision to close down a factory in North Wales. Bernard Wardle was, as a *Financial Times* headline put it, "going nowhere fast".

Taylor received little practical help from Ferguson Lacey. "He was always very nice and kind," he says, "but if I said we have a cash flow problem, his solution would be to buy General Motors."

Through the summer, Taylor had to hold creditors at bay, and desperately see what could be salvaged from the company. He arranged to pay off the debts in instalments, while chopping out 1,500 jobs, closing factories, and slashing low margin products. The headquarters building in Knutsford was disposed of, complete with a staff of thirty. Taylor chose three managers who were already at Wardle's – the finance, commercial and operations directors – and loaded work and responsibility on to them. In the six months to August, the company lost £2.5 million.

The next stage was to find something to make. Having reassembled the machine in Blackburn, the team decided to concentrate on Revotex's sound-deadening product, Dedpan: soon, Jaguar and Rolls-Royce agreed to use it in their car doors. Another product that had been on the back burner – a low-cost leather cloth for the backs of car seats – was also rushed into production. Bernard Wardle inched back from the brink.

Thirty-seven year old Tim Hearley also found himself in charge of a shattered motor supply company. But, where Taylor managed himself out of a hole, Hearley attempted to deal himself out. Since he had married the boss's daughter in the 1960s, he had kept a close eye on Coventry Hood and Sidescreen. An Oxford physics graduate, he had not become directly involved, preferring to work first in the City as a chemical industry analyst and then as a consultant, to build up his business skills. The family company, meanwhile, was churning out car seat covers and hoods: it employed 600 people in Bedworth and Birmingham. In 1977, Hearley had used his influence as non-executive director to bully the board into diversifying by buying Beaver Group, which made building chemicals, paint and foam for furniture.

In 1980, Michael Edwardes announced that he was closing MG, Coventry Hood's major customer. Hearley, acting as the family representative on the board, took action. He happened to meet Alan Curtis, chairman of Aston Martin, who was trying to raise the money to take MG over and keep production going. He offered to help, and found himself pumping £450,000 into the prestige car firm, giving Coventry Hood a 21 per cent share and himself a seat on the board. The quid pro quo was, of course, that CH would be able to go on supplying MG's hoods.

Throughout spring and early summer, Curtis and he scampered round the City trying to raise £11 million. Edwardes had promised to keep the Abingdon works open until July 1. They failed, and MG closed. But Coventry Hood and Hearley had been sent off on a quite different, and interesting path.

By mid-summer the terrible twins, British Steel and British Leyland, were in bad trouble again, pulverised by the recession. BL was back down to a miserable 15 per cent share of the collapsed UK market. In the first six months of the year, its car sales in the USA had slumped by 50 per cent, with every model being sold at a loss. Following a series of law suits against the company, and Ralph Nader's accusation that they were unsafe, it had had to discount to sell: the TR7, which cost £6,000 in Britain, was going for less than £4,000 in America. With massive redundancy payments and launch costs of two cars (the Ital and Metro), BL was already looking stretched – it had drawn down the bulk of the £300 million handed over by Sir Keith before Christmas. Now, with the market flat as a pancake, a vicious price-cutting war was biting into profits. Only the advertising men were profiting as knocking ads filled the papers – a sure sign that things were getting bad.

Although BSC was making surprisingly good progress at agreeing redundancies with the unions, its outgoing chairman Sir Charles Villiers announced in July that the corporation would not, as he had previously thought, be able to meet its cash limits. It would overrun by £400 million. Joseph accused him of lacking urgency in seeking a cure. Sir Charles was hurt. "When we are making 52,000 people redundant, it's a fantastic suggestion to say that we are not determined enough," he said. Joseph remained sceptical but, on July 26, relented: BSC would be propped up with unquantified funds.

In July, Fodens, the Cheshire-based heavy lorry maker, called in the receiver: 2,500 jobs were on the line. An Economist Intelligence Unit report predicted a massive downturn in commercial vehicle production, with the UK falling furthest: 9 per cent in a year.

Businessmen were becoming increasingly outspoken in their criticism of the government. "I can live with either a strong pound or high interest rates," said the chairman of one major company. "But not both." Another asked: "Does the government understand the real nature of the damage it is doing to industry? I doubt it increasingly."

As summer wore on, the government was finding it more and more difficult to square its philosophy with grim reality. Sir Keith was criticised for his lack of decisiveness. Would BL be taken away from the NEB? Would the NEB sell off its shareholding in Ferranti (now successfully turned round)? Would the Meriden Motor Cycle Cooperative have its debts written off? The minister just could not make up his mind.

Where he had taken action, it was actually to give in, as with BSC. At the end of July, Mrs Thatcher spoke proudly in the Commons of £6 million that had just been handed out to troubled Dunlop, as well as another £25 million given to Inmos. Sir Keith backed down on Harland & Wolff, the Belfast shipyard, offering £66.5 million over two years, and indicated that he accepted that British Shipbuilders would need fresh aid. He put a brave face on his crumbling free marketism. "There has only been a delay in meeting the targets. The objective has not been changed," he said.

When Bowater announced that it was going to close its Ellesmere Port newsprint mill, throwing 1,600 people out of work and halving Britain's newsprint capacity, Mrs Thatcher invited both management and unions to Downing Street. She offered the company money (which was rejected) under Heath's 1972 Industry Act; it smacked powerfully of the old beer-and-sandwich days of 1970s corporatism.

Was this part of the new phrase, "constructive interventionism" that was being bandied about Whitehall? It was made clear that the government was now more prepared to back research, to help companies get products to the market, and to use its own purchasing clout to help firms develop new products.

And how much was Joseph influenced by a new report produced by a former DTI adviser, Alan Peacock? It compared British and German state aid, and pointed out that the Germans used aid selectively to run with market forces, not try to counter them. So, when British Leyland was being propped up in 1974, Volkswagen was allowed to close the factories it wanted: instead the government subsidised the formation of new businesses in affected areas. And German support for the computer industry between 1971 and 1975 was shown to be four times higher than British; it was pumped solely into research and development, while British aid almost all went into the "flagship", ICL.

But it was difficult to tell if there really had been a change of policy at the Department of Industry, which was paralysed by the indecision of Sir Keith.

At the end of July, as Sebastian Coe and Steve Ovett were proving at the Moscow Olympics that the British need not lose at everything, the first piece of economic good news came through. Inflation had fallen from 21 to 17 per cent. Much of this was due to the running-out of the VAT increases that had been imposed a year before, but the underlying trend was encouraging. About the same time, minimum lending rate eased by 1 per cent under market pressure.

Joseph popped up from behind his parapet to announce a new initiative. He had always allowed that the government could involve itself in helping companies start up. Now, seven decaying inner city areas were declared Enterprise Zones. For ten years, they would be freed from red tape and rates.

In August, bad became worse. Manufacturing output was 9 per cent lower than a year before. Throughout Britain, 250,000 unsold cars were parked on wasteland and airfields: deep discounts meant that no one in the car business was making money. The 1,200 workers at Talbot engine plant in Coventry were working full time – on Thursdays. The government was spending £10 million a month subsidising workers on short time, to keep them off the jobless register.

Three quarters of the domestic appliance industry was on short time. Jobs were even being shed at so-called recession-proof industries: Courage and Ind Coope, the brewers, announced 810 job losses (they had not been helped by summer weather that had matched the economy). In the third week of August alone, 20,000 redundancies were announced, covering an ever-wider spectrum of industry. Firestone sacked 1,000, ending its tyre production in Britain. Reed Paper axed 700 jobs, Clark's Shoes 500, Massey Ferguson tractors 680, Hoover 440, Revlon cosmetics 100, Thorn Electronics 500. And so on. The next week Courtaulds announced the closure of seven mills: it blamed the strength of sterling. By the end of the month, the score had reached 35,337 – 3,512 in textiles, 18,553 in engineering and steel, and 13,272 in other industries. And these were just notified redundancies: if fewer than ten people were thrown out of work at a time, they did not reach the score card.

In the last week of August unemployment went through the 2 million mark – for the first time since the 1930s.

Even Establishment commentators were starting to wonder whether it would be better to prop firms up rather than hand out unemployment benefit. The government should recognise, wrote Anatole Kaletsky in the *Financial Times*, "that many a closure – even of a firm that is somewhat unprofitable – is a loss, not a gain, to the British economy".

Sir Keith's concessions, where he made them, were piecemeal, not wholesale. When the CBI asked for £145 million of regional grants to be brought forward, he said no. The Confederation had backed the government the previous year when it introduced a four-month payment delay on regional development grants. Now, it had changed its mind, pointing out bleakly that payments would fall anyway as industrial investment declined.

An unnerving feature of the jobs massacre was the meekness with which workers accepted their fates. "I have never known Midland workers so docile," said Phil Povey, a Midlands AUEW official. "They had been punched and battered by the flood of redundancies and short-time working. There is a feeling of despair and resignation – that they must let this government, much as they detest the policies, run its course."[2]

Men laid off by the big firms like BL and BSC had been given large redundancy payments to cushion the shock; Lucas offered £1,200 above the legal minimum when it made 3,000 electrical workers redundant, and was flooded by volunteers offering to take the money. But even those who got the minimum were disinclined to fight too hard. Unions did not seem to blame employers, or even the government. "We certainly don't hold the company responsible," said Duncan Simpson, AUEW convenor at Talbot's one-day-a-week plant. "To an extent they are victims of circumstances."[3] The speed of industrial collapse had, it seemed, shocked them into submission.

Something else distinguished this recession from other downturns. Managers were being thrown out of work at the same rate as shopfloor workers. In March, the Professional and Executive Register had about 60,000 people looking for jobs; by the end of the year, the number had risen to 117,000. The shock to the British executive was far more brutal than it would have been to, say, his American equivalent. For British corporations really did offer lifetime employment to people who wore suits. Often worse than the loss of salary (most managers were given decent compensation) was the loss of status – for company cars, executive toilets, and sheer prestige could be more powerful drivers than cash.

An unexpected trend gave some comfort to the government when trade figures for the first half of 1980 came through: manufactured export volumes were higher than they had been a year earlier. While car exports had slumped by 8 per cent, chemicals were up slightly, food and drink up 7 per cent, and intermediate manufactures – components of products finished abroad – were up by more than 12 per cent. The Treasury was perplexed: although it was normal for export earnings to rise initially when a currency hardened (because the same amount of exports would earn more in sterling terms), it could not think of any reason why volume should rise; indeed, the effect of rising sterling should be starting to reduce it.

There was a precise enough reason, though. Only the UK and the USA – the countries where interest rates had been pushed through the ceiling – were suffering from a deep malaise. The rest of Europe was feeling a little chilled, but other markets, especially wealthier-than-ever oil states, were still buying goods as if there were no tomorrow. In Venezuela, JCB found it could sell lots of excavators simply because the oil-rich locals would buy another as soon as their current one broke down. The Middle East and Nigeria were booming, and British managers' unrivalled skills at selling in difficult markets were being rewarded. The UK's export managers had never lost the frontier spirit that sent them hacking through jungles and trekking across deserts to sell their widgets. All they were doing now was hacking and trekking a little harder.

Three months after the recession had hit, it was possible to see more clearly what had caused it. The answer, predictably, was not simply that we were paying ourselves too much, as Joseph would have it. If that was true, why were sweet companies, which were scarcely bothered by international competition, laying workers off? And why did a Philips factory that made music centres have to close when labour costs were only 30 per cent of overheads?

The truth was that high interest and exchange rates were destroying operations that might otherwise have survived. But the long-term decline in profits, which had rendered them vulnerable, had more complex origins.

Exactly where those origins lay depended on the industry. There were two broad categories: those that were genuinely affected by external forces, and those that were paying the price of their own mistakes. In the former category came textiles, shipbuilding, low tech electronics: their problems blew in from the East, and the main criticism of management was, sadly, that factories were not run down fast enough.

Engineering's problems were, by contrast, mainly of its own making: all that sterling and borrowing costs did was to make them starkly and uncomfortably clear. The real problem was something nearer Joseph's analysis: British products were not good enough, and cost too much.

In between were industries with a foot in each category. The chemical industry had management problems, but it was also suffering from worldwide overcapacity. Even British Steel's overcapacity could not be blamed entirely on incompetent managers, blundering ministers or intransigent unions: the world steel market had collapsed. And the poor old sweet people had their internal problems – but none so great as the VAT increase imposed by Chancellor Howe in 1979. Children, it seemed, had extremely high elasticities of demand: when the price of sweets went up, they stopped buying them.

On September 1, Vickers merged with Rolls-Royce Motor Cars. It was seen as a defensive link-up to protect both companies against the ravages of recession. Vickers, once one of the great names in ships and aircraft, had been stripped of both by nationalisation in 1977. Now its main activities were in engineering, tanks and lithographic plates; its reputation was similar to that of many companies: poor profit record, ineffective management. Rolls-Royce, which had been floated as a new company in 1973 and whose management was well thought of, had been spending furiously on its new model, the Silver Spirit, which was due for launch in October. Vickers had the cash, from its nationalisation compensation, Rolls-Royce had the brains and a nice new product; the two came together, with very little rationalisation.

It was becoming clear that a new mechanism was at work in the economy. Sir Keith Joseph reckoned it was the managers' job to get tough with workers and bring settlements down. Instead, unemployment was doing the job for them. At the TUC conference in Brighton, there was what ministers chose to call "an outbreak of realism" as delegates called for moderate pay demands.

In the weeks before the conference, Lucas workers had voted to accept 10 per cent, a third of their original claim, and the rest of the motor industry looked likely to settle for around the same. It was more than just an outbreak of realism, of course: it was an outbreak of fear.

Inflation was now coming down, not yet because of lower wage rises, but because industry, desperate to keep ticking over, was slashing prices. Company profits were squeezed further. The City, which had remained sunnily optimistic throughout the summer, was stricken when it saw solid GKN cutting its interim dividend for the first time since 1935. Its chairman, Trevor Holdsworth, said that profitability in the UK was zero. Throughout the engineering sector, sharply lower profits were being reported: Babcock & Wilcox, down from £15.4 million to £6.1 million, Bridon from £8 million to £5.7 million. More alarmingly ICI, bellwether of the economy, saw its second quarter profits falling by 55 per cent.

Even more depressingly, ICL, the computer giant, sacked 140 graduates three weeks before they were due to start work. It blamed the "deteriorating economic situation", and said it would pay them one month's salary – £350 – in compensation.

If any complacency was left in industry, it was shattered when ICI's third quarter results were announced on October 23. After the record first quarter, the giant blue chip went into the red for the first time in its history. To stop City speculation, chairman Sir Maurice Hodgson had told his accountants to produce the figures five weeks early. Not only was there a £10 million loss, but write-offs and provisions for closures came to £150 million. The City was being offered total profits for the year of less than £250 million, against the £600 million it had forecast.

Hodgson, a shrewd, soft-spoken myope, knew that his lumbering giant was vulnerable. Although some divisions had been working to cut costs since the 1960s, ICI kept its reputation for being good on technology, paternalistic – and hopelessly overmanned. "The thing about the 1970s was that sometimes ICI did very well, sometimes not so well," says current chairman Sir Denys Henderson. "But recessions and losses were things that happened to other companies."

At a meeting in 1978, Hodgson had identified the problem areas: utilisation of working capital, energy efficiency, capital productivity, and overmanning. In the fourth quarter of 1980, says Henderson, "we realised this exercise of Hodgson's had practical implications. That was when the whole ICI machine swung into action."

Orders went out to all ICI's units to run for cash, pulling in working capital. They would not, they were told, get any capital expenditure unless there was a very good reason. The worst affected businesses were fibres, petrochemicals, plastics and dyestuffs. The losses, Hodgson explained, were largely due to the recession in the UK, where 40 per cent of sales were still made. The closure of more than 100 British textile mills had slashed the customer base of several divisions, and the market was too feeble to allow higher raw material prices to be clawed back with higher selling prices. Strong sterling, meanwhile, had stopped ICI making up the difference overseas. Hodgson said he wanted a bracing climate in terms of the value of sterling. "But there is a difference between that and freezing to death."

There was also a structural reason why the chemical industry was clobbered so suddenly. After the first oil shock, there had been a boom in the industry, with newly rich oil states all building colossal plants. These were starting to come on stream at the beginning of the 1980s, just when the market was turning down: they generated severe overcapacity. Chemical plants everywhere had to cut production levels, pushing them quickly into loss. Hodgson admitted that ICI had been slow to react: it had gone on investing when it should have been cutting. Now, the lumbering elephant had to move like a gazelle.

## Recession

The Tory conference in October was as cheery as *Macbeth*. Mrs Thatcher declared she was "not for turning" on economic policies, and while Sir Keith Joseph claimed to bring a message of hope, his solutions were depressingly long-term. Interest rates and inflation would come down, "sooner or later" and it would take "some years of understanding by management, unions and wage earners to become really competitive again". No one seemed convinced. He was, not surprisingly, booed off a British Aerospace site at Brough by its workers. But managers were becoming less and less convinced by Josephism. "Competent, proven companies are now finding it impossible to compete in world markets," said CBI president Sir Raymond Pennock.

At the 1979 CBI conference in November in Birmingham, employers had given a rapturous welcome to the new government. A year later at Brighton, its director general Sir Terence Beckett, ex-head of Ford UK, told the conference that the government did not really understand business. "We've got to take our gloves off and we're in a bare knuckle fight on some of the things we've got to do," he said. Immediately five companies resigned from the Confederation, but Beckett was putting in extreme form what many businessmen felt: that the government's insistence on holding up interest rates was going to kill British industry before it cured it. There should, most thought, be at least a 3 per cent drop, to 13 per cent.

The pound, they believed, must also come down: Sir Michael Edwardes declared that if the government had not the wit or imagination to use North Sea oil to help industry's competitiveness, it would be better to "leave the bloody stuff in the ground". "Companies are gasping for their lives," said Leslie Fletcher, chairman of Glynwed. "And facing the prospect of their last gasp far more quickly than anyone could have anticipated." Statistics showed that many had taken that gasp: company liquidations in the first nine months of the year were 56 per cent up on the previous year's level.

Ironically, just before the conference, the pound reached its peak, hitting $2.454 on November 4. The government thought that it was its "monetary continence" that had pushed it there; others thought it was North Sea oil. Most likely, it was a mixture.

And in the week the conference was held, the FT Industrial Ordinary Index sprinted 20 points. The City was looking ahead – and thought it could see things industry could not.

Managers were not slow to grasp the one advantage they could from the recession. Unemployment could not only keep wages down, it could be used to quell all sorts of unrest. Ford, which had been hit by 254 disputes in the first 198 days of the year – mostly at strife-torn Halewood – introduced a new disciplinary code that would dock £18 from the weekly pay packet of wildcat strikers. Sir Michael Edwardes, role model of the new macho managers, was busy trying to impose 6.8 per cent, the third single-figure rise in three years, after BL had lost £180 million in the first six months of 1980.

There was a surge of militancy at Longbridge, home of the new Metro, and at Cowley, by workers who felt they had been restrained long enough; compared with the Ford workers, they had indeed been put upon. On the night of November 21, workers stormed a factory building at Longbridge, breaking windows and damaging cars. Frustrations had built up after 500 people had been laid off because the company had run out of seats for the Metro.

The future of BL was at great risk. Joseph showed his displeasure by raising the borrowing limit by only £1 million, against the £1 billion Edwardes was asking for. Rubbing the message in, Rolls-Royce, high tech and for the moment looking healthy, had its limit raised from £750 million to £2 billion to allow it to develop new aero engines.

In a last desperate attempt to show that they still had teeth, leaders of Britain's steelworkers, railwaymen and miners formed a new triple alliance to resist any further closures at British Steel. "No one union alone can take on the government," Bill Sirs declared. In fact, Joseph was being comparatively gentle with BSC: at the end of November he gave it another £110 million.

At the end of November, Courtaulds confirmed what everyone knew – that the textile industry had got it in the neck. After the misery of sacking 25,000 workers in a year, it announced interim profits down from £30 million to less than £3 million. Those would have been a lot worse had it not been for the astonishing speed with which the 44-year-old chairman, Christopher Hogg, had moved. Even though volume dropped by 15 per cent, manpower was trimmed as rapidly, so that output per man remained static.

In the mid-1960s, Courtaulds had expanded into manmade fibres and fabrics, spending heavily to make itself internationally competitive. Then, in the 1970s, Far Eastern countries got hold of machinery that was just as sophisticated, added it to their low labour costs – and swept markets away. Japanese and American money had funded massive investments in ultra low-cost countries like Thailand and the Philippines; the efficient British retail system made it easy for imports to penetrate the home market. The pressures were already formidable when the strength of sterling tipped Courtaulds' textile divisions over the edge. But unlike other textile bosses, who reacted with knee-jerk cuts to the recession, Hogg, a Harvard business school graduate who had taken over in 1979, tried to be more ordered: he delegated as hard as he could, and left it to unit managers to decide what to do with their businesses. They did everything to save cash, and Courtaulds balance sheet remained undamaged.

Metal Box, the big can maker, halved its interim dividend at the end of November. Chairman Denis Allport confirmed that the fog of the steel strike had lingered long. It was not until August, he said, that managers had seen the extent to which it had disguised what was happening to the market.

Then ICL, glamour company of the 1970s, announced profits of £25 million – half the level expected a couple of years back. The usual reasons – high sterling, high interest rates, high inflation – were given. Old industries disappearing down their own smokestacks was one thing. But if Britain's major high tech company was in trouble, what chance did anyone else have?

On November 21, Howe announced a mini-budget. On the one hand, public spending was being squeezed even further, especially in defence and social security. On the other, £250 million more would be spent on job creation and training schemes in 1981–2. And, most important for business, minimum lending rate would be cut to 14 per cent. At last, both interest and exchange rates were on their way down.

In early December, Ian MacGregor told union leaders that there would be no pay rise at all in January – at least for six months. He also said another 20,000 jobs would have to go. A year before, a 2 per cent offer had triggered a thirteen-week strike. But MacGregor's plans were better than the unions had been hoping for, and there was hardly a flutter of rebellion.

It had been quite a year.

# 6

# *Hangover*

The recession rolled on for two more years. Unemployment headed inexorably upwards, reaching 3 million in January 1982, flattening out at something above that. (Statistical S-bends by the government were to make the total difficult to follow.)

During the first half of 1981, manufacturing output continued to fall. Attention switched away from devastated factories. To the Iranian hostages, freed in January. To the inner cities, which erupted into riot in April, then again in July. To assassination attempts on President Reagan and the Pope. And, on a jollier note, to the Royal Wedding at the end of July.

But if newspapers had decided that industry was last year's news, its problems were still at the dead centre of the political stage. Thatcher and Joseph stood behind Chancellor Howe as brickbats were hurled at him demanding that he counter, rather than encourage, the recessionary forces.

Critics drew links between the riots and unemployment, and pointed to the widening North–South divide as the South–East remained impervious to the economic downpour. One-industry towns that had been devastated were all outside the South-East: Consett had the highest unemployment rate in mainland Britain (although several textile-based towns in Northern Ireland easily topped its 26 per cent figure). In February, car making in Scotland ended when Peugeot announced it would close its Linwood plant (built during the Macmillan anti-Midlands days), adding to what a union leader called the "museum belt" of large closed factories in west central Scotland.

In the absence of a coherent formal opposition (the Labour Party was riven with strife, the SDP was not founded until March), it was left to Tories to take Thatcherism on. Much of the criticism centred on the lack of an "industrial strategy", or positive government plan to help industry. The roles of the interventionist French and Japanese governments were repeatedly emphasised. Everyone pointed to North Sea oil revenue, which was now pouring into the chancellor's coffers and which, the critics said, was being frittered away on supporting the jobless, rather than creating jobs. At a Cabinet meeting in July, Mrs Thatcher had to fight off a mass attack of wet wolves demanding a U-turn.

The March budget was aggressively deflationary. Although interest rates were brought down by two points, taxes on cigarettes, alcohol and tobacco were increased sharply. Public spending was squeezed where it could be, even though the sharp rise in social security payouts forced the total higher.

The 1981 budget came to be seen either as Thatcherism's finest hour, or its profoundest folly. The greatest depression since the 1930s had been deliberately deepened, apparently in the name of an unproven ideology. At the time, the folly brigade were much in evidence. Just after the budget, 364 eminent economists published a letter in *The Times* saying that the government's belief that deflating demand would bring inflation under control and automatically cause a recovery in output had "no basis in economic theory". Roger Eglin, the *Sunday Times*' astute industrial editor, commented wryly that that probably signalled the turning point in economic decline. "If anything can be more wrong than one economist, it is 364."

But one thing that the recession had done was to throw into high relief weaknesses and strengths that had been invisible, or ignorable, before.

It was not until the half-year results for Lucas came through at the end of January 1981 that its managers realised just how bad things were, even though they had made a formidable number of redundancies over the previous nine months. The group had made the first loss in its 100-year history and, Tony Gill says, "it had a shattering effect on management".

Failures and failings popped up in the most unexpected places. Many high tech companies weathered the recession well. In February 1981, 1,000 jobs had even been created to make Clive Sinclair's new mini-television in Dundee, while the rapid growth in the latest buzz industry, information technology, kept many small companies comfortable enough. But parts of high tech were now becoming "mature", and as vulnerable to recession as any metalbasher.

Despite ominous signs in previous months, there was much amazement when in early 1981 a new lame duck waddled out of the reeds of high technology. It was called ICL. Only in Britain could a company at the leading edge of technology develop severe arthritis when it was less than fifteen years old. ICL, a child of the Wilson IRC merger boom, was a big company, with big government links. It supplied most of the state's computers, its policy was influenced by the needs of the state and, until 1979, it had been 25 per cent owned by the state.

Although large by British standards, and the largest computer company in Europe, its turnover was about the same as IBM's research and development budget. Yet it had continued to produce a range of computers comparable with the American giant's. In particular, it had tried to keep up in the mainframe market, where development costs were astronomical and which was, in any case, starting to stagnate as attention shifted to smaller machines. When IBM began a price-cutting campaign on mainframes in 1979, ICL desperately repackaged and downgraded products to cut its own prices. Thrown-in features became expensive extras, and customers started to resist.

In the five months to February 1981, ICL had lost £20 million, and was imploring the government to underwrite its £60 million research and development budget. The immediate causes of the profits collapse were familiar enough: recession at home, and exports slammed by high sterling; it was hit particularly by the new thriftiness of its main customers in Whitehall. But the fact that ICL was as vulnerable to these as any old smokestack industry showed just how unperceptive its management had been. It had, in fact, shown exactly the same symptoms as many British failures: inability to change, over-reliance on the government, weak commercial strategy. As Rolls-Royce had found out ten years before, technical achievement alone was not enough.

Questions were asked about whether ICL should be saved. It would not be big enough to take on IBM head on, yet it lacked the nimbleness to fill niches. If, as in Germany, the government restricted itself to funding research and development, would not other companies spring up in its place? At the end of March, Sir Keith gave way: ICL would have £200 million in government guarantees. In exchange, a new management team headed by Christopher Laidlaw and Robb Wilmot was installed. Immediately, they announced 5,200 redundancies and a wage freeze.

*

## Hangover

The recession had shown up strengths, too. Lucas's loss was entirely due to the collapse in the motor industry. But its results would have been a lot worse had it not been for attempts to diversify into the aerospace business, which yielded a £7 million profit. That, crucially, allowed it to keep up its research and development spending.

Other firms had similarly been sustained by their more robust arms. A number owed their bacon to South African subsidiaries. If there is a recession, the price of the old fall-back, gold, starts to rise and with it the fortunes of its main producer. So, while Glynwed's British operations were languishing, with Leisure cookers losing £600,000 in the last half of 1980, its South African kitchen appliance company waded in with profits up from £3.2 million to £8.5 million. Similarly, Metal Box made 65 per cent of its meagre 1980 earnings from South Africa. Profiting depressingly from this seesaw as it tipped South Africa's way were dealers who travelled from auction to auction, picking up perfectly good but redundant plant and sending it south.

With the oil price high, OPEC countries were the saviours of other firms. Leyland Vehicles, which had opened a brand new factory right in June 1980, kept going by selling to oil-rich markets: it supplied big bus fleets to both Baghdad and Teheran. A whole raft of smaller companies was kept afloat by buoyant Middle Eastern markets: they might not be earning much from them, but they were at least keeping the factories open.

No one, not even Sir Keith Joseph, had foreseen how deep the recession would be. With a handful of astute exceptions, managers had been caught standing on one leg, and not a very robust leg at that. Their reaction had been to cut, cut and cut again. There was nothing very scientific about it. But although trimming was to continue for at least two years, the wholesale butchery was coming to an end. What was left?

The whole structure of British industry had been changed. Manufacturing output was some 14 per cent lower than it had been in 1979 when it bottomed out in the summer of 1981: in motor vehicles, the fall was 29 per cent, in textiles and clothing 24 per cent. Job losses since 1970 showed how swathes of industry had been thinned: 89,000 in cars, 37,000 in shipbuilding, 262,000 in textiles. The Lancashire textile industry had lost half its 109,000 workers and during 1980 alone, 200 medium to large clothing factories had been closed. What old-style industry was left was having to become more efficient while high tech industry was, with the exception of ICL, doing very nicely.

In other words, there had been a shift from old to new industry, and from labour- to capital-intensive industry: plausibly, the results of a successful industrial strategy. But nearly all this was done by the recession and by North Sea oil. Sir Keith admitted this: "We didn't foresee that sterling would rise so high." In typically self-critical style, he also admitted that the mechanism that he foresaw would bring wages down had not worked. "We had assumed that to the extent that sterling rose, management would convince unions that unit labour costs had to be reduced; to achieve that, either wage increases had to be modest or negative, and/or productivity had to jump. In fact, in many cases, managements failed to persuade unions and people lost jobs in swarms. You can blame management or unions, or both." But the job of controlling "unit labour costs" was done nevertheless – by the terror of unemployment. It raised an interesting question: had the world been economically healthy when the Tories came to power, could they have engineered the economic pressures on business that Joseph saw as vital? Everyone knew the artificial Barber Boom did not work. Would a Howe Slump have been any more sustainable?

The most hopeful sign of all was nothing to do with Thatcherite policies. If anyone could be thanked, it was the Wilson government of the 1960s.

The root cause of the British malaise was the lack of professional management. But, through the 1960s and 1970s, a new type of manager had been climbing discreetly up corporate ladders throughout Britain. By 1980, they were close to the top and a few, including John Egan of Jaguar and Christopher Hogg of Courtaulds, had just scrambled into the hot seat. Most were products of the new (technical, not plateglass) universities, probably grammar school educated, more than likely from a modest background. They were ambitious, and they rejected – despised even – the old values. All they could see was that Britain was going down the pan and that nobody was doing much to stop it.

There were, brutally, two benefits from the job losses themselves. First, masses of indirect workers were shed. These were support staff: working on maintenance, in the stores or accounts. Here was an area where British industry had become hopelessly flabby. Second, for the first time, managers were being sacked as fast as shop floor workers. A lot of good managers went, of course, but the majority who lost their jobs were older, old-style people. The new boys were allowed to take their place.

Constantine Folkes was an unusual sort of new boy. He took over as chairman of John Folkes Hefo in the summer of 1981. At twenty-eight, he was the youngest chairman of a quoted British company. Eight generations back, in 1699, Josiah Folkes had set up a farrier's business by a stream in Lee, in the heart of the Black Country. Folkes' headquarters was still yards away from that first "factory", although it now controlled thirty-five subsidiaries, in industries that ranged from fitted kitchens, via heavy foundries to property; the family had become a major industrial landowner over the centuries. The Folkes family controlled the group – Con himself owned 17.5 per cent of it – while other shareholdings were spread throughout the great and the good in the region. The group, which turned over £70 million in 1980, was a typical, if above averagely successful, Black Country company.

Con Folkes had done his apprenticeship by working in all areas of the group, then by training as an accountant. He reckoned he knew its strengths and weaknesses as well as anyone, and knew too that its weaknesses outnumbered its strengths many times over. A dapper young man with just a trace of a Midlands accent, he had a genuine interest in business: perhaps because the company was so old-established, the Folkes family had long passed through its dilettante phase.

He also knew that Folkes had survived the recession solely because of its property interests. The rest of the group had to be restructured, and he waded into the task. First job was to sell five Rolls-Royces, unacceptable symbols of wealth at a time of cutbacks. Then he closed three businesses, announced that the company had gone into loss, cut the dividend, restructured the rest of the companies and froze managers' salaries – all within his first three months. But his attempt to instil a sense of urgency into his managers was, he discovered, a depressing exercise.

"I was amazed that men couldn't accept that they had to get into the office at eight, not nine thirty, and stop having three-hour lunches," he says. "Even if it meant getting the sack, it seemed that that was a risk they were prepared to take."

Throughout industry, the management balance was being tilted towards the new boys. Tony Gill saw a crusty layer of management in Lucas, not at the very top, but in the next layer down. They were old-school managers, who revered the amateur and believed that the group owed them a living. As these people were cleared out, younger managers could come in and get down to things they had been itching to do for years.

One of the first lessons that was learnt was that executives would have to take their share of the cuts. Dining rooms were closed, managers started to travel second-class. It was easy to see whether a company had new managers or old: you just had to count the executive toilets.

The executive war cry had been, "Let managers manage!" Now,

nothing could be simpler. Companies that were teetering on the brink, in areas of high unemployment, found their workforces had learnt bovine docility. The speed with which the unions had lost power was dramatic; it owed little to the new Employment Act, and a lot to simple fear.

Managers took grim pleasure from their new authority. In the past, said the director of a Birmingham engineering company, "the power of the unions was so great there was total confusion all the time – even in the most mundane activity. Now we get total cooperation in spite of the tremendous cutbacks we have had to make. Will the new mood endure? Well, you still get union leaders saying 'you wait, our day will come', but the redundancies have got rid of the troublemakers who came to power in the 1960s. We've taken out thirty of them like they were a cancer."[1]

The device pioneered by Edwardes in his "Blue Newspaper", talking directly to the workforce, was used to bypass the unions. Soon, the 1970s debates about whether managers should be allowed to talk directly to the shop floor sounded ridiculous, even to unionists.

Not that peace broke out overnight. But by far the most sustained strikes in 1981 were in the public sector, and particularly in the civil service.

British Leyland workers held out longer than most against an increasingly hard-line management. The company started 1981 with a strike over the sacking of workers who had taken part in the Longbridge "riot", and continued in the autumn of that year, when the company responded to a 20 per cent wage demand with a 3.8 per cent offer. Edwardes said he would sack all striking workers and close the affected plants. He then softened his macho tactics by offering a little more money. The shop stewards voted to reject it by 238 to 12 but proved themselves out of touch with their increasingly desperate members: at mass meetings, the workers accepted the offer and the strike fizzled out after a couple of days.

Just as public sector workers, who felt less threatened by redundancy, were more inclined to strike, so companies that had survived the recession in best shape found it hardest to pacify their workforce. Ford was trying hard to introduce Edwardesian working practices, but was suffering from its own relative success. Helped by the new Escort, it had survived the recession well despite continuing industrial relations problems at Halewood, and was starting to look further ahead. It launched its AJ – After Japan – programme, which would combine investment in robots with greater labour productivity. Ford told the unions it wanted to shed 10 per cent of its 55,000 manual workers every year for the next four years and had introduced Japanese-style "quality circles" at some of its smaller plants. These groups of foremen and workers were supposed to meet regularly to talk about ways of improving productivity and quality, and to solve minor problems.

But Ford still had many of the practices that Edwardes had managed to suppress. Productivity was half that in the Belgian or German plants, partly because of the persistence of archaic practices. For example, ten men were employed to drive cars from testing bay to parking area; Ford wanted the testers to drive the cars themselves. And "tip dressers" were employed to keep the tip of welding machines clean: Ford said that was a job the machine operators could do themselves.

The unions considered much of the AJ plan to be a crude way of cracking their power. One aim of quality circles was that they would make workers feel more involved in their work; the flip side of that was that they would tend to emasculate the influence of shop stewards. Without the dreadful threats that BL could wave at its workers, managers found it difficult to impose their will. Industrial strife rolled on.

Even shorn of its industrial relations problems, the Steel and Leyland show made good viewing, with the government dipping into the taxpayer's pocket whenever a new crisis arose. In January 1981, Sir Keith Joseph announced another £990 million was being given to British Leyland, taking the total the group had absorbed since 1975 to £2.3 billion: it had lost £535 million in 1980. He said the group had done "just enough", in terms of introducing new working practices, reducing overmanning and launching the new Metro on time, to earn the money. The Metro, at least, was a success, and by early 1981 was Britain's best-selling car.

Edwardes' latest plan showed that the group could start making money in 1983. But there was a lot of hard work to do first. The UK truck market had collapsed from 80,000 in 1979 to 61,000 in 1980 and 45,000 in 1981. Although Leyland Trucks had opened a brand new factory in Lancashire in October 1980, it had yet to produce a new range of vehicles, and was losing share in a collapsing market. The important Nigerian market disappeared in political and economic chaos, and the group plunged £35 million into the red in the first six months of 1981. By May, Jaguar was losing £2 million a month as John Egan struggled to bring some quality back into the cars. That month, too, Edwardes announced that TR7 production would cease, killing off the last mass production British sports car, and that the Solihull factory, opened only in 1976, would close down.

In February, Joseph agreed to give BSC £880 million over 15 months to support MacGregor's survival plan. His junior minister, Norman Tebbit, called British Steel "this squawling baby. We can't give it away. We can't even leave it on the doorstep."

In fact, though ministers did not know it, the baby was responding well to its new American nanny. Ian MacGregor was no dummy: he was a professional.

MacGregor's plan marked a turning point, for it reversed the steady decline in production. Steel production in 1980 was 11.2 million tonnes (against 28 million tonnes a decade earlier), but he intended to build an industry with a capacity of 14.4 million tonnes. The effect on morale was immediate.

The American himself was having a big impact. His rough-hewn personality, deep knowledge of the industry, and habit of wandering around the shop floor talking to workers went down well. As soon as he had started work in July 1980, he had begun a reorganisation of the management to decentralise and try to make each plant a profit centre. Like Edwardes at BL, he understood the damaging effect of bundling units together into an impersonal monolith.

Perhaps even more important, MacGregor worked well with Bob Scholey, who had been chief executive since 1973. Universally known as Black Bob, because of his helmet and sharp tongue, Scholey had been passed over repeatedly for the top job. He had been criticised for his handling of the strike, and had then managed to get on the wrong side of Mrs Thatcher by refusing to regard its result as a triumph. His main enemies, though, were civil servants, who were uneasy with his bluntness and utter lack of suavity – even his moustache was lopsided. But he and MacGregor had much the same attitude to BSC's problems, and soon developed mutual respect.

The fact that MacGregor had set his "Alamein line" at 14.4 million tonnes astonished those who had nicknamed him Mac the Knife. Because of the ructions caused over his appointment, he was allowed to do what he wanted with minimum government interference. He also persuaded Eurofer, the European steel cartel that shared out markets, to give BSC some of the market back that it had lost in the strike, by threatening to dump cheap steel on the Continent. "The idea got around that since I was a new boy I might well be completely unpredictable," he wrote, "so that even the Germans felt it was worthwhile trying to humour the British."[2]

## Hangover

MacGregor's main surprise was that he intended to keep the South Welsh plants open, if vastly slimmed down. At the end of 1980, there were 5,701 workers at Port Talbot, against 12,584 in 1970. But it was at Llanwern that the shock was most effective. Built on a greenfield site in the early 1960s, it had never lived up to its promise as a model plant. Dispute after dispute had dogged it and after the strike, the workers of Llanwern felt their fate had been sealed. "All the time people in the pubs, everywhere, were telling us exactly when we would close," said works director Bill Harrison.[3]

Under this threat, changes in working practices were pushed through. A multi-union committee was created, and demarcation lines were scrubbed out. Clerical workers would take over operatives' jobs when needed, production operators would drive fork lift trucks, and fitters and electricians would swap jobs. A year after the end of the strike, BSC was claiming to have reached European output levels. Each man at Llanwern was taking 4.6 hours to produce a tonne of steel, against eight hours in 1979.

The speed of the turnaround owed much to the fact that the investment that had been made in the 1970s had never fulfilled its potential because of what MacGregor called "picayune little fights". Now, everything suddenly came spectacularly right as management and workers started to pull in more or less the same direction. The foundation was solid: it was just a matter of getting the walls straight.

Already, BSC was making its first investments. Massive continuous casting machines, which would produce steel in a continuous strip rather than in ingots, were being built at Scunthorpe, Port Talbot and Stocksbridge. When these were installed, BSC would be able to produce steel that was as cheap and high quality as the Japanese.

In early 1981, there were few obvious straws for British economy-watchers to clutch at. Redundancies were being announced at virtually the same rate as in 1980; in February, the destruction of the toy industry continued with the collapse of Airfix.

A possible straw was the government's invitation to Nissan to build a plant in Britain. Bids poured into Tokyo from 100 local authorities, all desperate for the jobs that would be created. The leader of the West Midlands council was among those to make a pilgrimage to Japan. There had been grim satisfaction in Coventry and Birmingham as British Leyland and Peugeot had closed their plants in Merseyside and Scotland, putting an end to the despised "M6 production line". The Midlands had been deliberately starved of jobs as the Macmillan government had directed investment elsewhere; now the Midlanders felt they deserved some of those jobs back. Unemployment had risen from 85,000 to 170,000 during 1980 but, because of its past prosperity, it did not qualify for regional aid. Surely the Japanese would see the sense in putting a car factory where car factories belonged?

It was a happy thought, but a mistaken one. It soon became clear that the last thing the Japanese wanted was truck with the old, dreaded and unionised motor workers: the shortlist was Wales, Humberside and the North-East.

The March budget, which had been so unpopular with industry, did contain some small but significant straws. In particular, there were moves to encourage entrepreneurs. Sir Keith Joseph's great complaint was that Britain needed more entrepreneurs, and his solution was to create a healthy business climate in which they would flourish. That was proving tricky for the moment, so the government weighed in with direct help.

A business start-up scheme was launched, which gave tax relief to anyone investing between £1,000 and £10,000 in a new company (later it became the Business Expansion Scheme). The National Enterprise Board was to have a subsidiary that could make low-interest loans of up to £50,000. And, after much lobbying, the government decided to set up a loan guarantee scheme, which gave an 80 per cent government guarantee to banks lending money to small companies. This was the most important change, and its introduction reflected badly on banks' support for industry. Although a commission, chaired by Harold Wilson and set up in 1977, had recently cleared the banks of culpability, there was a widespread feeling that they did little to help industry – particularly smaller firms.

The comparative uselessness of banks to companies went back a long way. In the early nineteenth century, a uniquely British division had been established as City families (Barings, Rothschilds et alia) had melded into the old aristocracy while industrialists were still building their satanic mills. The "formal" break came in 1878, after the collapse of the City of Glasgow Bank. The banking establishment decided it wanted as little as possible to do with industry, and shied away from links that were growing ever stronger in Germany. There were exceptions – the Midland Bank grew mighty on its funding of industry – but the City of London stuck to its core role of financing trade and projects round the world. When Britain's role as a trading centre slipped away in the 1970s, the City jumped gratefully on to another non-industrial bandwagon – the euromarkets, which specialised in channelling surplus oil funds to grateful but bewildered third world governments.

\*

# HANGOVER

Mark I'Anson had a little trouble with the banks. He was the sort of person Sir Keith Joseph was looking for, a man who was prepared to have a go at starting his own business and, better still, to do so in the town with the highest unemployment in mainland Britain. In 1981 he was 26, and had just arrived in Consett. The demolition squad had started work on the steelworks, it was raining, and gangs of men were standing around on street corners. "It was awful, like a scene from a Depression film," he says. But he was looking for somewhere to set up, it had to be in the North-East, and he was swept off his feet by the bouncy man from BSC Industry, John Hamilton. BSC Industry had been set up in 1975 to cushion the effect of the coming cutbacks in steel, and in melancholy Consett it was about the only ray of hope there was. I'Anson had been trailing round the councils around Newcastle but no one had impressed him as much as Hamilton: "He was the only bloke I could understand," he says.

His ignorance of business jargon was not surprising. A teacher's son and mathematical whizzkid from Wakefield, he had gone from grammar school to Selwyn College, Cambridge, and found himself in his mid-20s working as a computer researcher at the Open University. He and a colleague, David Liddell, were convinced that a new sort of microprocessor, the Motorola 64000, had enormous commercial potential, but government stringencies meant that they could take the development no further within the university. They decided to have a go on their own, found a company in Newcastle that had the expensive software they needed and, forming a joint venture with two of its partners, started to look for a base in the North-East.

Hamilton sat I'Anson down, taught him what a balance sheet was, and found him a local accountant. Then they started looking for finance, and discovered one reason why life was so hard for entrepreneurs. Venture capitalists, such as there were, were interested only in bigger projects: the banks were not interested at all. Only when the recently announced loan guarantee scheme was waved in front of them would they take any notice, and they were not gracious in their support. As well as charging 3 per cent above normal commercial rates, they demanded the directors' houses as security for the 20 per cent of the £75,000 loan not covered by the government. But it was a start, and I'Anson started to design his computer board: his research and development budget was modest at this stage – enough to pay for scissors and a great deal of coloured sticky-backed tape to mark out the circuits.

\*

Industry was not impressed by the government's decision in mid-1981 to slash university funding, which had been regarded as untouchable since the Wilson days. The government thought it was putting pressure on the airy fairy plateglass universities, whose graduates were finding most difficulty in getting jobs. The idea was to get academics, who were notoriously unwilling to treat with industry, to look around for commercial sources of funds. Somehow, though, the cutting formula put some of the heaviest burdens on the most "useful" institutions. Salford, which had been upgraded from a technology college in the 1960s and which had traditional links with industry, had its grant cut by 44 per cent over three years.

In May, the National Enterprise Board finally disappeared, joining up with the National Research Development Corporation to create the British Technology Group. The idea was that it would concentrate on funding high tech projects.

As early as the spring of 1981, some commentators were starting to challenge the received opinion that Britain's industrial base had been wiped out. They maintained that what the recession had done more than anything was to point up managerial deficiencies, and force change where it otherwise would not have happened. As if to back them up, output finally stopped falling.

In the West Midlands, companies that had been working three-day weeks were creeping back up to four days as the period of destocking came to an end. Birmingham Chamber of Commerce and Industry reported a slight pick-up in optimism in mid-March. Any company that had survived the recession more or less intact was breathing a huge sigh of relief, and wondering what to do next.

In May, Midlands industrialists told the CBI that their export orders were starting to improve as the pound softened (it was now at $1.92, although still high against European currencies). British Aerospace, which had been partly privatised in February, announced that its orders for the brand new 146 airliner had broken the half billion pound barrier.

The first statistical signs came through when industrial production was seen to have risen by 1 per cent that February – the first rise in nearly two years. There was still plenty of bad news, though, as the recession rippled through the economy: consumer spending was falling and investment, which was 6 per cent below 1979 levels, was still on its way down. Through summer and autumn of 1981, business surveys showed that output was bumping along the bottom, rather than making a concerted effort to climb away from it. This was more of a bottoming-out than a recovery: destocking had stopped, but restocking had yet to begin. Already, though, there were signs of troubles being stored up for the future; even while unemployment levels reached new heights, some engineering firms were complaining of a shortage of suitably skilled workers.

By January 1981, Brian Taylor had re-established an operating profit at Bernard Wardle. He had decided it would be a good idea to rationalise British PVC production by combining with a Turner and Newall subsidiary, Storeys Industrial Products. He had first approached Turner and Newall at the end of 1980 and had received "a bit of a dusty answer". But in the middle of 1981, T & N's own recession-induced problems were so bad (it was to go into loss in the first half of 1982) that its management contacted Taylor, and the scheme was revived.

Suddenly, Taylor found he had to learn the ways of the City. For Storeys was in an even worse state than Wardle's, and a complete financial restructuring was clearly needed. Ferguson Lacey, now labelled as nothing but a jetsetting wheeler-dealer by the City, wanted to get out – it emerged that his master company, NCC Energy, was on the point of collapse. So Taylor and his team decided to buy Wardle's themselves, and then take over Storeys. It was a bold decision, but it could, if it worked, give him the independence he had always craved.

In September Sir Keith Joseph left his job at the Department of Industry to take over Education. "Few industrialists will mourn his removal," said the *Financial Times*. "While some have recognised his honesty and intellectual approach, most have become frustrated by his refusal to face up to their detailed problems. Businessmen believed that his insensitivity to them prevented him and the Prime Minister adjusting policies to prevent too much permanent damage to the country's industrial base."[4]

By October, ICL was back in the black. The City and government had been calmed by the 59-year-old public school and Cambridge chairman, Christopher Laidlaw, while 36-year-old whizzkid Robb Wilmot had rushed around creating necessary turmoil. A child of the new generation (Worcester Grammar and Nottingham University), he had spent his working life in the American hothouse environment of Texas Instruments. Now, working 100 hours a week, he had spent a frantic six months cutting costs (a quarter of the 30,000 staff went, as well as most of the top management), introducing new products and making a series of arrangements with foreign companies – one was with Fujitsu to use its microchip technology, another with Mitel, a Canadian company, to move into telecommunications.

The other high tech creature of the state, Inmos, was having a tougher time under its new owner, the British Technology Group. The planned South Wales plant had been held up by government dillydallying, and some directors were arguing that it would be better to restrict production to Colorado, where it employed 680 people: 80 per cent of production went to the States, nearly all the rest to Japan. Plans for Inmos to become a major British chip producer were looking a little faded.

On October 5, a new anti-ulcer drug, Zantac, was launched by Glaxo, the "quoted university" that had earned itself a reputation as being good on research but weak on marketing. A City analyst had written in 1980: "It is difficult to escape the belief that Glaxo is one of the more somnolent of the major pharmaceutical groups." Suddenly, the City realised that it had not been sleeping, just beavering away among the test tubes without making much of a fuss. When analysts realised that Zantac did the same as Tagamet, which had been responsible for a quadrupling in the American SmithKline Corporation's profits over five years, Glaxo suddenly became a glamour stock.

In mid-November, British Leyland signed a long-term agreement with Honda. As the Triumph Acclaim, no more than rebadged Honda, had just been launched in Britain (it was selling astonishingly well), it seemed strange that Ray Horrocks, BL's car group chairman, should call the link-up a "true partnership of equals". But that, more or less, was what it was, for Honda as a car producer was also small by world standards, and its engineers had developed considerable respect for their British counterparts. Honda had never produced a big car; BL was expert, but had no funds: hence the agreement to design a new executive vehicle.

The Japanese must have swallowed hard, though, for they signed the agreement just after 2,200 workers at Longbridge had walked out over plans to trim their tea breaks, in exchange for a shortening of the working week. It was the sort of typical British dispute that foreigners always found hard to understand.

There was not much joy in company results from 1981. What better figures there were – such as GKN's £35 million profit – came from cost-cutting rather than any upswing in demand. By the end of the year, though, there were some signs of improvement. In January 1982, the CBI industrial trends survey showed a net positive balance of 8 per cent in companies' optimism about the business situation. In other words, slightly more thought it would get better than worse. No one, however, could possibly claim that the outlook was sunny.

# 7

# *Hiccuping to recovery*

Nineteen eighty-two was the year when business receded even further from the headlines. What space the Falklands conflict did not take was used up by various public sector disputes, especially on the railways and in the health service. British Leyland had a couple of strikes: one was at the troubled commercial vehicles division, the other over pay. This last was settled with workers accepting a two-year pay deal – a device that was starting to emerge as inflation headed downwards and the unions became more pliable. It was also the year when business had less to complain about. Sterling was on its way down, interest rates tumbled, to 9 per cent in November, and in March inflation reached 5.4 per cent, the lowest rate since 1970.

In January 1982, Jaguar made its first profit for many moons. It owed its success to the USA, where sales had suddenly taken off. It was not that Jaguars had suddenly become remarkable: they had just stopped breaking down. In the recent past, dealers would regularly sell a car on a Friday and expect to see the enraged customer on Monday morning demanding his money back.

This recovery was particularly significant. Not only did it give a revered British institution much of its lost pride back, it showed what a little bit of professionalism, as opposed to panic management, could do. When John Egan, a 40-year-old product of Imperial College and the London Business School, had arrived in April 1980, he had conducted clinics to find out exactly why the cars were going wrong, and set up teams to cure the faults. More than half were due to poor-quality supplies, and Egan came down hard on the offending subcontractors, forcing them to pay replacement and dealer labour costs for any fault they had caused. The company then took a close look at BMW, Mercedes and the Japanese car makers and as a result introduced the "Pursuit of Perfection" campaign: it was based on what had been learnt in Japan. The cars' problems, 150 of them, were listed, and task forces were set up to tackle them; the directors took on the twelve most serious. In 1977, a quality-based propaganda campaign had been run, without much effect. But it had been followed by a series of "video nasties", and by the time Egan introduced quality circles, the workforce was sufficiently aware of the problems to accept them without a murmur.

The previous September, Egan had brought his American dealers over to show them the 1982 model. They were impressed, but said they could sell only 3,000 in the year. He persuaded them they could manage 6,000 – and in the event, they sold 10,500. He had shown that a little attention to detail could have a remarkable effect.

Just as managers lifted their heads above the parapet, and started sniffing the pleasant whiff of lifting demand, a secondary recession came blowing in across the Atlantic. If the full force of recession had slammed into Britain in 1980, it had rather skirted other economies by. America had slowed down a good deal, but had a long way to go before it stopped completely. Other industrialised countries had been hit only in specific sectors, the oil states were still living it up on their $30 a barrel-plus earnings, and Latin American countries that had had vast loans forced on them in the 1970s were still busy spending them.

Now, in 1982, the American economy was screeching to a halt, braking the rest of the world with it. Most of the blame was put on the US administration. It was running a massive budget deficit at the same time as strangling money supply: the result, inevitably, was sky-high interest rates. The falling price of oil, and the results of gross overborrowing were pushing previously healthy markets like Brazil and Mexico towards bankruptcy. Even the Arab oil states were having to cut back on the speed at which they were building new and unnecessary airports.

An immediate casualty of the weak American economy was John De Lorean's sports car factory in Belfast. Set up with £18 million of taxpayer's money in 1978, De Lorean, a flamboyant ex General Motors boss, had said he hoped to sell 20,000 of the stainless steel gullwinged creatures. But, said the receiver when he took over in February 1982, 8,500 would have been a more realistic target. Sales had never lifted off and the government finally pulled the plug by refusing to put in more cash. It was the end of an ill-considered piece of state intervention.

By July, the CBI's optimism balance was back to minus 22 per cent. The view from the boardroom was hardly improved when Barclays Bank announced that it had 200 companies in special care, with more being added all the time. In September, Midland Bank said that it had between 70 and 80 companies with "Mexican style debts" coming to £350 million. Standing close behind the banks and playing an important, if near invisible role, was the Bank of England. It helped arrange financial reconstructions or rescue mergers for scores of companies. Usually, it let the clearing banks take the lead but on several occasions it played an active role – even, where necessary, delicately ousting a chief executive.

A new bogeyman, unfair trade barriers in overseas markets, became a big issue. Why should Spanish Ford Fiestas be allowed into Britain with a 4.3 per cent tariff, when Metros going the other way were lumbered with 36.7 per cent? Should we not do something about the Japanese habit of holding up Jaguars because they did not like the symbol on the headlamp switch? Perhaps we should be following the cynical French, who were channelling all video recorders through a tiny customs post at Poitiers?

There was indeed much to complain about, but the complaints were made mainly to express industry's increasing frustration that it was stagnating while what little extra demand there was was sated by foreign suppliers. The volume of imports had risen by 40 per cent since 1977, with Far Eastern makers knocking out their enfeebled British rivals. In 1982, probably for the first time in peacetime, Britain imported more manufactured goods than it exported.

British Steel was blown off course. After coming close to breaking even in March, it was losing £1 million a day by late summer, and MacGregor was again talking about closures. The reason was simple lack of demand, particularly abroad: capacity utilisation fell from 95 per cent to 63 per cent between July and October. Despite MacGregor's 14.4 million tonne Alamein Line, it looked as though only 10 million tonnes were needed. Foreign steelmakers had not reduced capacity at anything like the rate BSC had and, as their own markets fell away, they were inclined to offload cheaply wherever they could – including Britain.

MacGregor decided that the most marginal plant, Ravenscraig in Scotland, would have to be closed, and went to the new Industry Secretary, Patrick Jenkin, to tell him. To his surprise, he was told that this would make the government very unhappy. The Tories had been defeated in March at the Hillhead by-election by the newly formed SDP and was, MacGregor thought, going soft. "I think they had half an eye on the fact that we were less than a year away from a general election," he wrote.[1] He was probably right.

A boost to Britain's business machismo came in the middle of the Falklands conflict in May, when it was discovered that GEC might be about to rescue the ailing West German giant, AEG. Ever since Arnold Weinstock had merged GEC with AEI and English Electric in a virtuoso bit of dealmaking in the 1960s, his group had managed to remain relatively untroubled by the problems of the rest of industry. When the home market for power equipment collapsed in the early 1970s, GEC moved smoothly abroad, and won order after order in South Africa, Australia, Hong Kong. Weinstock believed in decentralised management: his headquarters, tucked away behind Park Lane, was absurdly nondescript for such a large company. By the early 1980s, GEC had generated a billion pound cash mountain, and the City was always on the look-out for clues that Weinstock might be about to spend it. The AEG deal came to nothing, but it was good for Britain's battered morale.

Less cheerful was the announcement that reality had finally caught up with the ultimate recession-proof product, the Rolls-Royce. Having announced at the beginning of the year that 1982 would be a record year, the car firm had to put 2,000 workers on short time in June. It had been fooled by the early upturn in sales after the launch of the Silver Spirit in 1980. Soon, however, it had become clear that company chairmen were becoming uneasy about buying a new car at the same time as laying workers off. Even if they decided to renew every other year instead of every year, after all, sales would halve. Then the US recession deepened, hitting the Rolls-inclined entrepreneurs over there, and sales slumped. By early June, 342 cars had been sold in America, against 460 in the same period the year before. British registrations were down from 666 to 417. The rationale of the Vickers/Rolls-Royce merger was looking a little shaky.

Despite the hurricanes and eddies whirling around the British and world economies, some managers were now starting to wonder what they would do when and if the recession was ever over. They had already made great changes, of course, simply by slashing their workforces. But you could only sack people once. What could then be done to keep productivity improving?

The government thought it knew the answer. It had declared 1982

Information Technology Year, and made it clear which way it thought industry should be travelling – towards high tech. Everyone was grasping at silicon straws. Articles began to appear along the line of "Automate or liquidate", and the only healthy sector in the economy was seen to be high tech, where Sinclair and Acorn were taking the computer world by storm. Figures were produced showing how few robots British companies had compared with everyone else – especially the Japanese. The unions did not like robots, because they were seen as a way of cutting unionised labour; the government liked them for the same reason, and the Department of Industry said it was prepared to subsidise their introduction. During 1982, the number of robots almost doubled, to 1,200, and at the end of that year, the first fully automated factory in Britain opened. Owned by the 600 Group, it manufactured a variety of engineering components on a production line controlled by a handful of whitecollar operators. The component was loaded automatically on to a conveyor belt, from where a robot arm would pick it up, present it to the appropriate tool for machining, then put it back on the belt to be picked up by the next robot. This was a Flexible Manufacturing System, and it would enable parts to be made in three days rather than three months, and by three men rather than thirty. The factory, which was mostly paid for by a £3 million grant from the government, was designed as a showpiece to persuade other companies of the virtues of FMSs.

Tallent Engineering in Newton Aycliffe, County Durham, had been making changes, but automation was not the half of it. In June, when the first Ford Sierra rolled down the production line, Tallent's chief executive Bernard Robinson breathed a sigh of relief. For he had banked the future of his company on a small part – the rear suspension arm – of the new car.

Robinson had been in charge of Tallent Engineering since the spring of 1980. A stocky, local man with a fondness for cigars, he had had to leave grammar school at 16 because his ex-miner father could not afford to let him stay on. He had joined Tallent in 1956 as an apprentice toolmaker on 34 shillings (£1.70) a week and had worked his way up through the production side. Robinson was a great learner. Early on he had been on a course on scientific management; and as a director in the late 1970s, he had taken himself off to night school to earn a formal accountancy qualification.

Tallent was a typical British company that had followed an erratic path from one activity to another. In the easy days after the war, it had pumped out cigarette cases, powder compacts and other "fancy goods". In 1955 it had been bought by Sir Charles Colston, ex-chairman of Hoover, who had used his jumbo £75,000 golden handshake to invest in what he thought would be the "appliance of the future", the dishwasher. He was wrong: sales of the only all-British machine never reached expectations as the public headed instead for automatic washing machines and freezers.

To keep busy, the Tallent factory also made paraffin heaters, until a domestic fire caused by a heater destroyed the market in the mid-1960s; and it built twin tub washing machines for John Bloom, before his Rolls Razor empire collapsed in 1964. As Bloom's biggest creditor, it took over his tooling and stock, and continued making various machines to the end of the 1970s. In this increasingly competitive market, it was in the unhappy position of being Britain's smallest manufacturer and, squeezed by other companies' price cutting, the Colston business was sold to Ariston. So Tallent switched again, this time to pressed metal fabrications; soon, its main customer was Peugeot Talbot, and particularly its factory at Linwood in Scotland.

Although the Linwood factory was under threat during 1980, it did at least keep operating, and Robinson could spend the summer working out a strategy for the company – the first time that had been done. He realised that Tallent's main weaknesses were its reliance on one customer, and its inability to offer anything that other firms could not offer just as well. The answer, he decided, was to make a small number of products, and make them better and cheaper than anyone else; then the biggest customers could no longer walk all over him. When Linwood closed in early 1981, two years earlier than he had expected, he had to accelerate his plans. Tallent's turnover halved to £1.5 million, and Robinson decided to take a risk.

He knew that the big car companies were starting to change their approach to subcontracting. For long, British suppliers had got away with producing shoddy and unreliable goods largely, it seemed, because car makers were prepared simply to put contracts out to tender and choose the cheapest bid. The only control was the crude one that the buyer could quickly shift to another supplier if things went wrong. The Japanese, they saw, took a different approach. Car companies not only gave long-term contracts to subcontractors, they also helped them to produce the best quality components. The relationship was based on support, not fear. There *was* a big British company that operated a similar system – it was called Marks & Spencer. But it was well out of the field of vision of the beleaguered men of the motor industry.

Jaguar was already moving down this road, and Ford was not far behind. It agreed to give Tallent a long-term order for rear suspension arms for the new Sierra, but only if it would invest £3.5 million – far more than its annual turnover – in new plant. British banks refused to come and see the factory; the money came from American and French institutions. In went Japanese robots, Swedish computer software and two massive 800 tonne presses (they at least were British) that would give Tallent an almost unrivalled ability to produce the components accurately and rapidly.

Not that Tallent's problems were over then. When this state-of-the-art equipment was switched on, it didn't work. Efficiency fell to 20 per cent, and Robinson struggled desperately to sort out the problems. Ford was not yet ready to lend a helping hand. Despite this, and despite having accumulated a frightening amount of debt, Tallent survived – thanks to the credibility the Ford contract had given it with the banks. The formula was a good one, and it would spread.

Over in the Manchester suburb of Blakely, Tony Rodgers was trying to put some life back into another sick child of the recession. After four years in Canada, he had just arrived back to take charge of ICI Organics.

It was not healthy. The begetter of much of ICI, with roots stretching back to early in the century, the group had made its money supplying dyes to the textile industry. Now it was wilting with its client base; it had been losing money since 1978, and there was even talk of closing the whole division down. Rodgers, a marketing man with more than a passing resemblance to Colonel Sanders (of the chicken), had spent much of his life in approved ICI fashion working in divisions round the world. He found he had taken over a group whose cash flow was haemorrhaging, and where morale had been shot to pieces: "I certainly felt an element of despair: there were a lot of extremely good people who had been fighting from the trenches for some time. My remit was, for God's sake, sort it out."

But first, he ordered the grass to be cut. It was growing three feet high all over the Blakely site, a result of a decision to cut every unnecessary overhead. But, Rodgers deemed, a boost to morale was worth the candle and the mowers moved in. Then, he sat down with his colleagues and tried to work out how to rescue the business.

In September, Sir Michael Edwardes left British Leyland. In his five years, he had cut 82,000 jobs, closed five major car plants, two truck plants and a bus factory, and killed off the MG. On six occasions, the existence of the group had hung in balance. He had never been popular with the government: British Leyland, in absorbing ever more money, had undermined its non-interventionist dogma more than anything else (later, Joseph said he wished he had had the courage to let it collapse). Yet Edwardes was, in the public mind, something of a hero, not only for the way he had taken on the unions and won, but for leaving the company in a condition that gave it at least a chance of survival.

Brian Taylor was making the transition from manager to dealmaker, by attempting a "management buyout" of Bernard Wardle. In 1977 there had been ten buyouts, but in 1981 there had been 124, with the concept accepted by the City and, with more or less reluctance, by companies. Buyouts were simply the purchase by managers of the company or subsidiary for which they worked. Finance came from investment funds or venture capital organisations, with the Industrial and Commercial Finance Corporation (later to be 3i), owned by the clearing banks and the Bank of England, dominating the business. The recession had encouraged MBOs, as slivers of companies were taken off on their own as an alternative to being closed down. Freed of big company control, managers who had escaped redundancy by the skin of their teeth tended to do remarkably well.

Taylor wanted to undertake one of the biggest buyouts to date – it would involve raising £5 million to give him the funds to absorb Storeys. His attempts to find backers taught him much about the City. "We were turned down by a dozen banks," he says. "The first thing they said was manufacturing industry, yuk! The second was the car industry, yuk! The third was plastics, double yuk! And the fourth was, look at that loss record. You might think that a management team that had done what we had done was worthy of attention, but they were interested in profit in the future, not glorious resuscitation. The fact that we could raise the dead was no proof that we could make the dead very healthy." In the end, it was personal chemistry that saved the day: Taylor hit it off with John Moulton, who ran Citicorp Venture Capital, and the bandwagon started to move: the buyout was completed in October 1982. "I suppose I was a bit perplexed at first that so many people didn't seem to think that what we had done was terribly impressive, but it was their money,' says Taylor philosophically.

In the autumn of 1982, IMP – Mark I'Anson and David Liddell's embryonic computer company – hired three men and started building computer boards; its factory was in the Consett steelworks' old counselling office. One of the three men was Jim Routledge. He had worked as a clerk for fifteen years at the works and had campaigned hard against the closure. The management had worked flat out to streamline the works, and the workforce had accepted the cuts: the factory had been about to come back into profit when it was closed, he believed.

After the Save the Works campaign had celebrated its defeat with a typically British parade, Consett had sunk into two years of phoney prosperity; redundancy payments were used to buy new cars or pay off mortgages, even as foreign companies came in to buy up the steelworks plant. The misery had been reinforced by the slashing of branch factories in surrounding towns: 3,600 jobs were lost at the steelworks; 3,000 more went in other companies in 1980 alone. In 1977, 12,900 people had been employed in manufacturing; in mid-1981, there were 4,400.

Routledge, 31 when the works closed, had decided to get himself retrained. At a Durham TOPS course, he had learned to be an electrical wireman; at the same time, he had sent off applications by the dozen and ended up with two shoeboxes full of rejection letters. He hated the idea of signing on the dole, and was considering moving to Wales when an assistant from the "Joke Centre" dropped in an application form that led to the IMP job. His pay was £5,000 a year, compared with £7,500 he had been earning at BSC, but that was the least of his worries. As he worked, building and testing power supplies, he watched the steelworks come down around him. "That was the worst bit," he says. "It was eerie."

In October 1982, another Employment Act was given the Royal Assent. In character with its author, Employment Secretary Norman Tebbit, it was a tough piece of union-bashing legislation, which had little of the caution of its 1980 forebear. Under the act, closed shops were further weakened, political strikes were made illegal and unions were made financially liable for unlawful action. The TUC had run a massive campaign against it, but there was little they could do, especially as most Liberal and SDP MPs refused to oppose it.

By the end of 1982, consumer spending was starting to pick up as inflation and interest rates fell. The final abolition of hire purchase controls earlier in the year had made personal borrowing easier, and the basis for a consumer-led upturn was laid. The housing market, which had been flat as a pancake since 1979, showed signs of life, leading to growing optimism in the building trade. As the home market became easier, companies proved that while they could sell abroad when they needed, they much preferred to pop round the corner.

Despite an increasingly favourable exchange rate, exports dropped by 1 per cent even as consumer goodies were sucked in.

Meanwhile, some interesting efforts were being made to pick up the shattered pieces the recession had left behind. The two state giants, British Steel and British Coal, both had their industrial regeneration companies, but some private sector firms were also making efforts to cushion the shock of wholesale closures. As part of its ad hoc approach ICI, for example, had managed to come to an arrangement with a much smaller company, to transfer a number of managers in their mid-50s. ICI paid their pensions, and the other company topped them up to full salary.

Rank Xerox was conducting two experiments. Its problems resulted only marginally from the recession. Rank, the cinema and entertainment group, had been wise enough in 1956 to contribute £600,000 to form a joint venture with a small American company called Haloid, which had developed the first plain paper copier. The new company, Rank Xerox, was allowed to make and sell the machines, Xeroxes, everywhere in the world outside the Americas.

Xerox technology was based on a series of patents, which made it impossible for any other company to copy the copiers and inevitably, both Xerox and Rank Xerox grew fat and slovenly on their monopoly (although equally inevitably, the Japanese affiliate Fuji Xerox, which Rank Xerox spawned, did not). When Henri Debuisser, Rank Xerox's personnel manager, first visited the largest European factory, at Mitcheldean in Gloucestershire, in the early 1970s, he thought the workers were on strike. "In fact," he says, "it was perfectly normal activity."

During the 1970s, the patents ran out, and Japanese competitors led by Canon poured in with better, more efficiently made machines. Although Rank Xerox never came close to making a loss, its profits started to drop sharply after peaking in 1977, and the management decided something must be done. The first stage was to shed labour, particularly at Mitcheldean: more than half its 10,000 workers would have to go. But the company was the only big employer in the area, and chief executive Roland Magnin decided he should do something to cushion the impact. So he allocated £200,000 to convert old mews buildings to a business park, with a series of workshops. Sharing many of the facilities of the factory, redundant workers were encouraged to start up on their own. At the same time, an intriguing scheme was set up to ease executives out of the company, while still benefiting from their skills. It was called Xanadu.

During the 1960s and 1970s, Rank Xerox had grown fast, and taken on hundreds of bright managers. As it slimmed down in the early

1980s, many of them became redundant but, rather than sacking them, Magnin decided to offer them a chance to start their own businesses. He told them that Rank Xerox would pay them their full salary for 100 days of consulting a year (reckoning that this was the amount of days' real work they got from full-timers), leaving them with another 120 working days to set up their own businesses. This arrangement would be guaranteed for three years, after which the company could decide whether to renew the contract. They were sold machines (including Xerox computers) and furniture at book value, and were given access to a pool of super-secretaries, called network coordinators.

A year after being set up in April 1982, Ian Roderick's Xanadu company Quantec Systems and Software had already turned over £100,000. Soon, there were 400 Xanadu businesses; there was a bias towards computing and consultancies, although they included a hotel, a bakery and a farm. The company was pleased too: it was able to clear and sell one complete office building in London.

During 1983, the economy started to gain momentum. There were still plenty of problems for the government, as lame ducks seemed to recover, then fall into apparently terminal relapses. In early 1983, it was the turn of British Shipbuilders to sink back. Its merchant yards, which had shed 23,000 jobs in the previous five years and which were mostly well equipped, were running out of work as the world recession put the brake on new orders. Worldwide capacity stood at 19 million tonnes, orders were only 6 million tonnes, and BS chairman Sir Robert Atkinson told his workforce in February that he did not think there would be any upturn for three years.

British Shipbuilders was always fundamentally the weakest of the ducks. There was no particular reason why Britain should not be able to keep up a healthy car, computer or steel industry; there was one why shipbuilding was unlikely to be profitable. As other European countries were finding, it was impossible to compete with the Koreans on price, and opportunities for filling specialised "niches" were limited. If it was any consolation, this was not just a British problem.

Meanwhile, at BSC, Ian MacGregor had been pondering what to do with Ravenscraig, following the government's veto on closure. In March, he thought he had the answer. He had learnt by chance that US Steel had problems with its Fairless works in Pennsylvania, which had obsolete open hearth furnaces. Ravenscraig, in contrast, had efficient basic oxygen furnaces that could produce continuous cast slabs. MacGregor thought that if Ravenscraig could feed Fairless with slabs, then at least part of both plants would be kept open. Fairless's job would be to turn the Scottish slabs into coil, sheet and tin. He reckoned 60 to 70 per cent of the Scottish workforce could be kept on, working in the hot metal end; about 2,000 jobs would be lost. His friend and counterpart at US Steel, David Roderick, agreed, and so did the chief American union leader. Unhappily, the union leader fell ill and in the ensuing confusion, a muddled account of the plan leaked out to the ISTC and the United Steel Workers of America. Soon, they were whipping each other into frenzies of discontent, and the plan was scuppered. MacGregor suspected the leak to the ISTC had come from somewhere in the Scottish civil service.

Ford's attempts at doing an Edwardes were looking fruitless. In March, the Halewood plant was back on strike over the dismissal of an assembly line operator for alleged vandalism. Although Halewood was one of the biggest employers left in a ravaged Merseyside, industrial disputes had still not been suppressed. In 1982, there had been seventy-three stoppages.

In February 1983, the Storeys deal was pushed through, and Brian Taylor started cost-cutting, concentrating all activities on to three main sites within six months. He closed down a factory at Lancaster, and moved his headquarters to a newer building at Manningtree in Essex. "The Storeys managers planned to close the Manningtree factory instead because they all lived near Lancaster, even though the factory there was a listed and totally impractical building," he says. Suddenly Taylor found himself with a strong, efficient group – and he owned 31 per cent of it himself.

The March 1983 Budget gave more help to small businesses. It included an extension of the Enterprise Allowance, which paid would-be entrepreneurs £40 a week while they found their feet. The Small Engineering Firms Investment Scheme was also boosted. This subsidised small companies wanting to buy new plant, and was designed to overcome lack of capital for investment in the latest machinery. It would, the government thought, be a good way of boosting the West Midlands.

In early 1983, British Leyland introduced the LC10, or Maestro. It was the car that Edwardes had wanted to bring in at the same time as the Metro, because it was bigger and could therefore earn bigger profits. It got good reviews, and the Cowley workers who made it showed their growing confidence by coming out on strike. The dispute was over a decision to abolish the right to stop work three minutes before the end of a shift – the so-called washing-up time. But the strike highlighted the same production line problem that had always dogged "Taylorist" operations: to the management, the few extra minutes of production meant another 100 cars a week; to the workers, it was another cruel and dehumanising adjustment.

By May, economists were agreed that the world recession was lifting – all indicators in the USA were pointing upwards – and British industry had a chance to taste some of the fruits of its abstinence. Profits were a quarter up on the year before, while consumer spending was encouraged by a budget that was verging on the expansionary. As the car industry had led Britain into recession so, much slimmed down, it was trying to pull it out. Car sales were 17 per cent up on the previous year, and manufacturers had thrown themselves into a vicious price cutting war to get the biggest chunk of the bumper August sales, when the new A-prefix would begin. With no hire purchase controls, dealers could force wonderful deals on customers with the same enthusiasm with which bankers had previously lent money to Mexico. Five-year repayments with £100 deposit were offered: not surprisingly, consumers jumped at them.

For the first time in the decade, most of Britain's economic indicators were pointing in the right direction. Inflation had dipped below 4 per cent, productivity was rising at 3 or 4 per cent a year, the fall in employment had finally stopped, even though demography was pushing the unemployment total higher. The UK was no longer the sick man of Europe, or the world, as other countries had been taking at least as great a pasting during 1982. In 1982, 12,000 companies failed in Britain, 20,000 in France, 25,000 in the United States, and even 17,000 in Japan. Satisfyingly for the monetaristic Tories, the French attempt at spending their way to growth had collapsed in despair. The Falklands were thanked, or blamed, for giving Mrs Thatcher her landslide election victory in June. But for those electors with jobs, rosier prospects played their part. Most encouraging for Tories was the trade unionists' vote: the proportion of unionists voting Labour was 12 per cent lower than in 1979.

British managers who had survived the first mad sort-out were able to sit down and wonder what had been done wrong before, and how they could avoid making the same mistakes again. For even if British industry was improving faster than others, it was from a far weaker state. The convalescence had begun, but would industry ever do more than hobble?

Just as managers were breathing a deep sigh of post-recessional relief, another threat arrived to disrupt their peace of mind. When in April 1983, the little-known conglomerate BTR laid siege to Thomas

## Hiccuping to Recovery

Tilling, the twentieth biggest company in Britain, managers knew that they were being put back on their mettle. For the stock market – reinforced by recovery on Wall Street – was setting off on a great bull market, and it was those who could ride the market best who would be able to control the rest of industry. To managers trying to get their factories into shape, they would prove something of a distraction.

# Part Three

## *The bull starts to roar*

# 8

# *Enter the predators*

Chaos is always a good time for clever people to make money. During the recession, there were plenty of opportunities for asset strippers to move in on enfeebled companies, break them up, and sell off property and machinery. But that, by and large, did not happen. Partly because the opprobrium that surrounded Slater and his like still hung in the air, but more because there were better ways of making money, there was very little gratuitous asset stripping.

What happened instead was that latter-day Jim Slaters saw that there were huge profits to be made in the very areas of "low tech" industry that many people had written off for good. There was, they thought, nothing wrong with most British companies that a little bit of tight management could not put right. They could make the assets "sweat", or produce more profits, even though the line between sweating and stripping was sometimes dangerously thin: occasionally, it was non-existent.

These were the predators, 1980s style. In tiny print first, then larger, names like Nigel Rudd, Greg Hutchings and David Abell started appearing in the pages of the financial press. They started up when everyone else was closing down and they epitomised one strand of 1980s manager – the bottom line superstar, the creature of figures, not fine engineering.

But they were not a new invention, indeed their two most powerful role models were busy saddling up to make the jump from being big to being vast. Hanson Trust and BTR jumped on to the back of the bull market with frightening aplomb.

When his family haulage business was nationalised in 1948, 26-year-old James Hanson took himself off to Canada, where he started another haulage firm. There he developed a style of management that was marked by caution: he would put cash in the left hand drawer of his desk, bills in the right hand drawer, and would never pay them unless the one covered the other. He was also keen on delegation, pushing decisions as far down his company as he could.

In the 1950s, he teamed up with Gordon White, whose father owned a printing company in Hull. To start with, the two were known more as young men about town (the ever-elegant Hanson was engaged to Audrey Hepburn for a year) than as business partners. In due course, though, White joined Hanson on the board of Oswald Tillotson, a Hanson family company that had escaped nationalisation. Then, the family interests of both men were merged with Wiles Group, which made sacks and fertilisers. In charge of a public company, and with a million pound loan from Lloyds Bank, they consciously set about creating the biggest industrial conglomerate in Britain. They were part of the set of high living young businessmen that included Slater and James Goldsmith, and they were found most nights gambling at clubs like Annabel's in London. They admired Slater, who was building his own industrial conglomerate and, between 1967 and 1973, bought a handful of smallish companies, mainly in building materials.

Like Slater, they believed that business was about numbers, and that it was indeed possible to make two and two equal five. The technique was to buy companies that were not well managed and to generate profits by stripping out the top layers of management, closing headquarters buildings, selling off subsidiaries that did not perform, and reducing labour wherever they could. They pushed responsibility down to divisional managers, and gave them rigorous financial targets.

Hanson (writing in 1986) explained that he preferred not to be thought of as a predator but rather as "a modern game warden using a tranquilliser dart to bring in an ailing animal, enraged and bellowing, for curative treatment... Experience has shown us that by their nature, old-established basic industries tend to have somewhat complacent and over-populated boardrooms, frequently unresponsive to change and therefore often outpaced by new products... Change from within can be a painfully slow business and in today's climate may simply take too long."[1]

Although there was no shortage of poorly managed companies in Britain, the political climate turned against conglomerates in the 1970s. It started with Heath's "unacceptable face of capitalism" speech, aimed at Lonrho, and continued with the new Wilson government.

The move by Slater away from industry, followed by his collapse, deepened the gloom; Hanson and White decided to concentrate on the other side of the Atlantic.

Establishing himself in New York, White spent $350 million on American companies between 1973 and 1980, mainly in the food and textile industries. (He even tried to persuade the Shah to give him $1 billion to invest in the depressed London and New York stock markets.) In Britain, Hanson got through only £56 million in the same period. Hanson Trust, as it had become in 1969, entered the 1980s with a turnover of £684 million: more than half its £40 million profit came from the USA.

In the 1960s, BTR was an entirely different creature. Founded in 1924 as the British subsidiary of the American Goodrich tyre company, it had expanded into other rubber industries, dropped the Goodrich connection and in 1934, become British Tyre and Rubber. In 1956, it stopped making tyres and by the early 1960s was producing a vast range of rubber bits and pieces with a workforce of about 15,000. It was a typically British, rather undynamic industrial company.

In 1966 the chairman, Sir Walter Worboys, chose a 41-year-old accountant called Owen Green to be his successor. Green, a Stockton-on-Tees grammar school boy, had commanded a gunboat in the war, trained as a chartered accountant and then become finance director of a hose manufacturer called Oil Feed Engineering. He came under the BTR umbrella when it bought Oil Feed in 1956, and worked his way around the country bringing life to bits of its sleepy empire. Worboys spotted his ability to motivate people and chose him for his job. In the year he took over, BTR lost £235,000 on sales of about £16 million.

Immediately, Green surrounded himself with a team of like-minded managers: John Cahill, Norman Ireland and Don Tapley. All except Cahill were accountants and they thought naturally in terms of figures: BTR soon developed a structure remarkably like Hanson's. That, he insists, was simply because it was best. "We have never bothered about how other companies were run. We have no role models." BTR tucked itself away in a tiny headquarters building in London's Pimlico, pushed control as far down the line as was possible, and kept control of otherwise autonomous companies through copious financial analysis.

The new team set about enlivening the business, buying here and there with an eye always on keeping a balance. First, with the encouragement of the IRC, BTR merged with the Leyland and Birmingham Rubber Company, doubling its size. Then it bought a motor component company to give it some exposure in the automotive sector.

Then it started to look abroad, buying companies in Australia, Germany and the USA. BTR was quick off the mark in tapping the North Sea bonanza: in 1974, it formed a joint venture company with an American firm to make modules and other equipment for offshore work. Unlike Hanson, who had his feet firmly planted in Britain and the States, but nowhere else, BTR moved all round the globe.

Green believed unequivocally that the singleminded pursuit of profit was the only thing that would make a company succeed. Like William Lever sixty years earlier, he thought that concentrating on other objectives, such as job creation, could only lead to disaster. But he was aware that that was not mainstream thinking at the time. "I used to make speeches saying that profit is not a dirty word," he says, "it is not like lust, it is healthy and lifegiving to our people." At times, later on, Green's dogged adherence to his own rigorous but unforgiving philosophy would get him into trouble. Even when a strike-bound South African subsidiary was coming under embarrassingly public scrutiny, he refused to show a softer side. Like Enoch Powell in politics, he would stick to his own intellectual guns, and damn everybody else.

The big event in the BTR year was the development of annual plans, which made managers understand exactly what they were supposed to do. These were monitored via detailed monthly analyses which poured into headquarters. In the high inflation days of the 1970s, price rises were ruthlessly passed through – inflation was not allowed to mask "real" cost increases. If a product could not sell at a sufficient profit, the BTR answer was to stop making it. Although the group's power was used to buy raw materials centrally, subsidiaries were not expected to buy from each other unless they gave the best deal. And, if a company wanted copies of the in-house newspaper, *Grapevine*, it would have to pay for them.

BTR was one of the few firms that knew that a recession was coming and cut back in anticipation: the workforce was trimmed steadily from 17,500 in 1977 to 12,350 in 1981. "Our planning process would show immediately that our labour content was out of kilter," Green says, "so we started whittling away in 1979." The question, he points out with justification, "is not why we were ahead of the game, but why others were behind". When the recession came, this control, combined with geographical spread, kept profits buoyant: even in 1980, its worst year, operating profit increased by 21 per cent.

By then, BTR was geographically well balanced and starting to expand outside rubber products: its range included a mass of specialist industrial widgets as well as rubber dinghies, squash balls and hot water bottles. Managers were encouraged to squeeze maximum profit out by concentrating on specialist or "niche" products where high margins could be commanded. BTR was large but not huge: the 122nd biggest industrial company in Britain.

The public first noticed Green when he was named Businessman of the Year in 1982. It was a rare break from cover: he did not appear in *Who's Who* until 1986. Not only did he dislike personality cults, he wanted nothing to do with the establishment. BTR would not join the CBI, which he despised as a tool of corporatism.

The big difference between Hanson and BTR was – and is – that Hanson's *raison d'être* was to buy and sell companies. Slimming down acquisitions was merely part of the dealing process. BTR would buy because it wanted a company that fitted in some way into the empire. Green was primarily a manager, not a dealmaker; profit growth would be generated as much by investing as by cutting.

Both companies came storming through the recession and quickly spotted that there must be bargain basement opportunities in the wreckage of British industry. Hanson's first big UK purchase was of Berec, maker of Ever Ready batteries, in 1982. BTR bought Serck, a valve manufacturer, in 1981.

Their big chance came when the stock market started off on its mad gallop. The bull market in London had actually been born eight years before, immediately after the stuffing had been knocked out of shares by the extraordinary events of 1974: in that year, the stock market lost half its value; on January 6, 1975 the FT30 share index sank to a low of 146. Then it started to move up, because the institutions, led by the Prudential, had decided things could not get any worse and had started buying. Others joined in and the market gained its own momentum: people bought shares because they thought they would go up, so up they went. By the time the Winter of Discontent came, the City would take no notice of any news but good news; 1980 was a good year for the London stock market.

In late 1982, Wall Street joined the fun, and the bull's muscles really started to ripple. As share prices zoomed ahead, the predators knew their time had come.

Early on April 5, 1983, Green's stockbrokers signalled the beginning of the great takeover boom by snapping up 6 per cent of the Thomas Tilling group, the twentieth biggest industrial company in Britain. Its business included Cornhill Insurance, Heinemann the publishers, Pretty Polly tights and a host of construction materials and engineering subsidiaries.

The tactics of the takeover game were developed by the players, using a rulebook drawn up by the authorities that laid down when a holding must be declared and when a full bid must be made. The predator's aim was simply to get investors to sell a majority of the target's shares to him, in exchange for either cash or his own shares. He would start by buying as many shares as he could (often in a "dawn raid", before anyone realised what was happening), then set about building his holding towards 50 per cent. There would be a deadline, which he could extend if he thought he was within grasp of victory. The target's aim was, of course, to stop him succeeding.

The structure of company ownership had shifted during the twentieth century. Before the First World War, most companies were private, owned by their managers or by the descendants of the founders. Between the wars, and more particularly during the 1950s and 1960s, this system broke down. Increasingly, families sold out their holdings, often to release capital to pay for death duties: the holdings were usually taken up by "institutions" – pension funds, life insurance companies or investment trusts who, with the government's national savings, were the main guardians of British personal savings; some were based in London, others in Edinburgh. Merchant banks in the City started to specialise in investment management, bundling together the pension contributions of scores of companies or local councils, and investing them in the equity (stock) market, as well as in property, government bonds or even cash.

The London stock market grew ever more important and sophisticated, while those in France and Germany, which were not fed by wealthy institutions, remained stunted and rather quaint. The British (and US) system had one major advantage – the institutions were rich enough to act as an almost bottomless source of new funds: these could be raised through "rights issues", by which new shares would be offered to existing shareholders.

Despite its size, the stock market was a genteel place right into the 1970s. Slater and the asset strippers had caused a stir, but after the 1975 crash, the City was in no mood to throw its weight around. Traditional stockbrokers, who were supposed to sell the shares, would still arrive on a rather late train from Kent or Surrey, make a few phone calls to contacts on behalf of their private clients, go out for a longish lunch, and then catch an earlyish train home.

During the 1970s, the City started to wake up. Analysts were employed by both stockbrokers and institutions, and company performance came to be monitored professionally. Even then, the job of an analyst would consist of producing weighty volumes, rather than instant news, and fund managers tended to leave their portfolios alone unless there was a crisis. Directors of public companies did not have to look constantly over their shoulders at their owners – which was one reason why they could pursue objectives other than profit. That started to change in the last couple of years of Labour rule, when the reaction of the financial markets began to have an impact on government policies, and the City's influence grew. At the same time, there was a gradual concentration of financial might in the hands of the fund managers; by the early 1980s, individual funds controlled up to 2 per cent of the entire London stock market. It became normal for the fate of the crucial 50 per cent of a company, even a large one, to be in the hands of fewer than a score of people.

Under the Thatcherites, who believed the market to be the supreme power, the City's authority was further bolstered. Their most important step was to remove exchange controls in late 1979. Suddenly fund managers started to measure the performance of their portfolios against American rather than British norms, and found UK performance wanting. There was now nothing to stop them transferring their entire portfolio overseas, and vast quantities of pension contributions and personal savings found their way into office blocks in Houston and the share register of General Motors. American-style monitoring groups were set up: these would rank funds according to their performance over the quarter or even the months, and fund managers did everything they could to stay near the top. If they did not, pension fund trustees – who kept a close eye on the rankings – might be tempted to transfer their money.

It was only when the takeover battles started that the weakness of the system showed up. Share prices would move up and down rapidly during these battles, and fund managers would always be tempted to sell out at a good price and take a profit, rather than give the target company a chance to manage itself back to health. They would find themselves the most popular people in town as executives and merchant bankers from each side would flatter, cajole and reason with them to accept their arguments. Personalities could be crucial. Typically, fund managers would be grammar school technocrats, probably actuaries; they would be faced by sharp-witted predators, most likely from grammar school too; by merchant bankers from the top of the public school drawer, probably very bright; and by the directors of the target company who would try to show that they too had been to the Clint Eastwood school of management. It was often a tough choice. Analysts and City journalists would pile in with their twopennyworth – their views would be noted. "Sentiment", the collective view of the City herd, would also play a part – if hostile takeovers were in favour, the fund manager would be more likely to support the predator.

The role of the pension fund trustee could be an anomalous one. He would normally be a manager of the company whose fund it was, and in that role would abhor the short-term shunting of stakes that the new climate encouraged. But as trustee, he was supposed to ensure that the fund got the best return, and had to insist that the juggling continue. It had been quite normal for a large chunk of a company's shares to be held by its own pension fund, a useful block to takeovers, but that was no longer true: now, the whole system thrived on the takeover game. Only the pawns, the companies involved, could lose.

Not that they had to. For directors of the prey, the experience of being hunted would be a chastening and toughening one. Apart from having to learn the techniques of the City in double quick time, they would also have to keep the interests of their shareholders at the front of their minds. Their prime duty was to their shareholders, not to the workforce or to anyone else. So, if a generous enough price was offered, they would have no choice but to recommend its acceptance – even if it meant putting themselves out of a job.

Sir Patrick Meaney, Tilling's chief executive, denounced BTR's raid, saying that any takeover bid would certainly be an asset-stripping operation. But it was a mighty battle: the old conglomerate, which had done enough buying in its time, fighting off the new, number-crunching aggressor.

Tilling was the company behind London's red buses. When they were nationalised after the war, it used the compensation to build up a conglomerate, growing particularly fast during the 1960s. When Meaney took over in 1973, he found a group that was running out of momentum in sluggish British markets, so decided to move overseas. Like Hanson and BTR, he spent heavily in the USA during the 1970s; that helped him through the 1980 recession.

But Tilling did not have the structure to control its overseas web, especially as many of its subsidiaries were in complex, high technology areas shunned by Hanson and BTR. It was also more gentlemanly in its approach to new subsidiaries. Where Hanson or Green would normally sack the directors, on the ground that it was their mistakes that had made the company vulnerable, Tilling would leave them in place and allow them to continue as before with minimal monitoring. When the US economy started to nosedive, Tilling's profits went down with it: slumping from £70 million in 1981 to £40 million the next year. Although its £2.2 billion turnover was more than treble BTR's, profits were about a third.

The takeover battle, the biggest ever staged in the City, brought the first wave of full-page knocking advertisements; each side denounced the other in ever franker terms. Sir Patrick Meaney, a stalwart of the CBI who operated from splendid Crewe House in Mayfair's Curzon Street, had to hurl his natural gentlemanly behaviour out of the window as the upstart from Pimlico closed on his company. The bid was, Sir Patrick said, "blatantly opportunistic"; his main point was that the profit downturn was cyclical, and that shareholders should hang on. Green riposted that a little bit of tight management could do a lot more than a cyclical upturn. Meaney's profit forecast for 1983, of £95 million, was "frankly incredible".

BTR pushed the price it was offering for Tilling shares higher and higher, but the real argument was a fundamental one: who would do a better job for the shareholders? Insults were traded, with aggression taking on paternalism: "There seems to be a womblike way of life at Crewe House" (Green), "They've done a fantastic job with a scruffy looking lot of companies" (Meaney). Financial journalists weighed in (they preferred BTR). And merchant banks and law firms rubbed their hands as their fees piled up. The tension mounted as BTR's June 8 deadline approached. Each day, its shareholding crept up, but the ace cards were in the hands of the institutional investors – the Prudential alone owned 2 per cent of Tilling.

At teatime on June 8, BTR announced it had acceptances for 58 per cent of Tilling's shares. The takeover marked the beginning of open season on British companies. Institutional shareholders, who had become increasingly important as holdings concentrated in their hands, were giving notice that they would no longer automatically back the incumbent managers. As industry had lost its gentlemanly air so, aided by bright young calculator-driven analysts, had the City.

If any old-style managers who had survived the recession thought they could relax, the Tilling result put them right. As Patrick Sergeant said in the *Daily Mail*, "the age of head offices like Crewe House is behind us: the poky boxes full of number crunchers are abroad in the land".

June 8, 1983 was a turning point: the first great victory of new over old. But what good, if any, would it do to British industry?

This, for BTR, was not only the leap into the big time, it was its first move towards becoming a true conglomerate. Nothing was sold, and turnover leaped from £725 million to £2 billion. That was the BTR style, and contrasted with Hanson's big deal of the year: he bought United Drapery Stores for £255 million, and immediately sold off assets worth £185 million.

In December 1981, C. Price, a private company, bought 51 per cent of W. Williams & Sons, a quoted foundry company based in Caerphilly, South Wales. W. Williams had had a bad recession, with losses reaching £996,000 on a £7 million turnover in 1981.

On November 5, 1982, the *Financial Times* announced that W. Williams, now described as having interests in BMW distributorships as well as foundries, had made a £3.1 million bid for Derby-based Ley's Foundries and Engineering, which had lost £2.4 million in the 1980–81 financial year.

An astute reader of the financial pages could have detected that something funny was going on. It was not, it turned out, so much funny as remarkable. For new boys were at work, trying to construct a 1980s-style conglomerate from the wreckage of the recession. Nigel Rudd and his partner Brian McGowan were, with a twenty-year handicap, trying to catch Green and Hanson up.

Nigel Rudd was born on the last day of 1946, the son of a weights and measures inspector and an accountant's secretary. He was brought up in Derby, where he went to grammar school and did well without being brilliant – good at maths, bad at languages. But his parents could not afford to let him stay on to do A-levels, so when he was 15 he went to work for an accountant friend of his father's. By the time he was 20, he had qualified, and took a job as company secretary in a local civil engineering company. When it was taken over by one of the many aspiring conglomerates of the late 1960s, London & Northern, Rudd found himself managing director of the civil engineers: he was 22. He quickly gained a reputation as a troubleshooter in the group, going into new acquisitions to sort them out, and often sacking people twice his age: his nickname was "the fireman". It was very practical business training.

London & Northern started to run out of steam in the early 1970s, and when it lumbered itself with a large housebuilder just as the property market collapsed, Rudd was given his biggest rescue task. Though its collapse could have brought the group down, he was left on his own and, doing everything he could "to keep the bankers at bay", he saved the company. By now, he had the urge to run his own business, and was in any case starting to fall out with his bosses. He persuaded them to give him, "as compensation for loss of office or as a thankyou", a housebuilding subsidiary in South Wales called

C. Price. It had a good land bank but was locked into a council house contract that was losing money. With a colleague, Roger Edwardes, Rudd paid bonuses to foremen to persuade them to work themselves swiftly out of a job. Within eighteen months, the two could sell the rest of the land bank, earning £1 million.

Rudd had not spent much time with his family – he had married when he was 22 – and now, at the age of 33, he decided to retire on the £600,000 share of his little bit of asset stripping. In the six months that British industry was being destroyed, he got his golf handicap down from 24 to 10, but also started to get more than a little fidgety.

It was in 1981, while watching *Magic Roundabout* with his 4-year-old son, that he finally decided that something must be done. He started looking for an unfashionable sector to invest his money in, and lighted naturally on engineering. He decided, as so many others had before him, to build an industrial conglomerate. Having worked for one, he thought he could avoid the pitfalls.

Rudd knew he needed a partner. He was a good dealer, good at spotting opportunities, and a good motivator. But he needed someone to knit the deals together and sell them to the City. The answer, he thought, was an old accountant friend from London & Northern days called Brian McGowan. He had been acquisitions manager for P & O and, Rudd considered, was "the best merchant banker never to have worked in a merchant bank". He was now working in Hong Kong as finance director of Sime Darby. Rudd flew there, persuaded McGowan, and flew back.

He then started to look around for a quoted company that he could use as a vehicle. Through his contacts in South Wales, he came upon W. Williams. It had done everything wrong, re-equipping heavily just as the automotive industry was nearing the edge of the cliff, and the family that controlled it was desperately trying to sell.

With £200,000 of his own money and £90,000 of McGowan's, his private company C. Price bought a bare majority of Williams: the rest of the shares were mostly held by a sceptical Welsh Development Agency. Even before McGowan had arrived back from Hong Kong, Rudd had closed two businesses to try to staunch the losses. He found Barclays Bank, perhaps sensitive about the reputation clearing banks had been earning, extraordinarily helpful: it added £1 million to his overdraft limit.

As they set about cutting the workforce, selling property and cutting stock levels to generate cash, Rudd and McGowan decided they had moved in at the right time. "We genuinely felt the thing had taken such a socking that we stood a chance of getting prices up, increasing margins and taking costs out to give the company a chance of surviving," says Rudd. They were not thinking about profit, but of cash. Like Hanson, they knew that if there was enough to pay the bills, then a company would survive: it was a fundamental concept that too few businessmen had understood.

As soon as Williams started to resuscitate, they turned their attention to their long-term goal. To buy other companies, they knew, they must have a share price that would give them muscle. The arithmetic of takeover by "paper", as opposed to cash, is simple. A share price is worked out more or less, as a multiple of profit per share; that multiple, the price/earnings ratio, is determined by the City's view of a company's future. A company that the City thinks is expanding will have a much higher P/E ratio than one that it thinks is going nowhere. If a predator is offering a target's shareholders shares in itself, it can clearly buy it more "cheaply" (in terms of the number of shares it has to hand over) if it has a high price/earnings ratio. As the acquisition will in itself push the profit up, the City scratches its head, and pushes the share price up again. It becomes, for the successful predator, a virtuous circle.

The virtue can be enhanced by a set of conjuring tricks known as "acquisition accounting", that are designed to improve the apparent profitability of a company after it has made a takeover. The commonest of these is to "write down" the value of stocks and assets of a company after it has been bought, then generate extra profit by selling them off. For example, the acquiring company can say that the inventory (stock) held by its new subsidiary has no value: this is applauded as "conservative accounting". But in the real world, the inventory does have a value, and when it is sold, the proceeds appear as pure profit: cash has apparently been conjured out of thin air. Furthermore, if the value of an asset has been written down, no depreciation need be allowed on it: as depreciation is one of the elements that reduces in the profit and loss account, profit will again be boosted. There are plenty of other similar devices, most of them quite legal, that can make an acquisition appear to be more profitable than it really is. The predators are masters of them.

Rudd and McGowan started to egg their P/E ratio, and thus share price, upwards. City folk are motivated either by fear or greed. The Williams pair appealed to greed, by selling their story to local financial journalists in the hope that the news would spread and investors would see Williams as a source of potential fortune. Here, they were lucky enough to find two men working on the *Western Mail* in Cardiff – one was the deputy City editor, John Jay; the other was Bill Jamieson, his editor (they both went on to become City editors on the nationals). Soon, stories about the young accountant who had come back from Hong Kong to rescue a Welsh foundry were being syndicated around local papers, and in due course the share price started to respond.

But until the share price was strong enough, Rudd and McGowan would have to use cash to buy other companies. They quickly persuaded Barclays to let them spend £200,000 on two BMW car dealerships in Hull and Doncaster. They thought that with a little careful management, car dealerships – especially upmarket ones – would be great generators of cash, as BMW paid for the cars in the showroom. They were right: they produced £300,000 in the first year.

The next deal, in November 1982, put them more firmly in the minds of the City. Rudd was keen to return to the Midlands and heard that a big foundry in his home town of Derby was in financial trouble. The family that controlled it wanted to sell. Ley's had been the biggest malleable iron foundry in Europe but was now losing money hand over fist. Rudd and McGowan reckoned they could use their experience to turn it round, and they could see it had assets of £11 million, made up of stock and money owed by debtors. Still they were not ready to use their paper, so they went to a bank that was prepared to lend them £3 million at a high interest rate. The family sold, half the business was closed down, land sold off – and within a year, Ley's was breaking even.

The 1982 figures for Williams Holdings, as the group became, were not on the face of it spectacular. It lost £315,000, and turnover was down slightly on the year before. But to those who knew – and there was an increasing number of them – Rudd and McGowan were on the way to great things. For just as they were ready to buy with paper, the stock market obligingly moved off on its great bull run. It was gripped by a fervour for takeovers – and Williams was only too happy to oblige.

As Rudd was sniffing around W. Williams in 1981, other bright young things were thinking along the same lines. They too saw easy profits rising up out of the murky swamps of low tech.

David Abell had risen to become one of Michael Edwardes' senior executives at British Leyland. He had made a fortune on the stock market in the late 1970s and was as well versed as anybody in the art of bringing finance and industry together. He bought as his "vehicle" a hairdressing equipment maker called Suter and, in 1980, left British Leyland. But he took part of it with him – Prestcold, the refrigeration division that Edwardes had been trying to sell. It was in a bureaucratic mess and Abell gave the management too long to sort itself out. This first acquisition was a false start. He lay low, and did not make another move until 1984.

Brothers Raschid and Osman Abdullah, educated at a home counties comprehensive and thoroughly British despite a name inherited from a Saudi great-grandfather (their grandfather came to England to set up the first Indian restaurant), bought into the cold roller and brass strip maker, Evered, in early 1981. In the previous two years, it had lost £794,000 and, like Rudd, they reckoned it could be turned round with some careful management. They had read the same management textbook, and they had the added advantage of some useful Middle Eastern backing. Their techniques were rather more successful than Williams' in the short term: in 1982 the company made a £370,000 profit.

In the middle of 1983, it was easy to miss Williams, Evered and Suter: their success ranged from modest to non-existent. But a few had spotted them, and thought them worth watching. So too was another little company, well away from the paralysed world of engineering. It was called Hillsdown Holdings.


When a country is at war, there are always areas where it is easy to forget it. In the industrial conflict of the early 1980s, swathes of industry remained untouched. Food production dipped slightly, as distributors cut their stocks, and then resumed its gentle upward path. The wallpaper maker Coloroll suffered a sharp drop in the summer of 1980 and another brief dip in mid-1982, but soon recovered as stocks were rebuilt – underlying demand was unchanged.

But these industries had their share of bad managers – probably more than their share. The cosy environment in which they lived meant that outdated structures and practices could carry on unhindered. Unhindered, that is, until bright young things came along and threw them into turmoil.

It was difficult to call Harry Solomon a bright young thing. He was, in 1980, 48, a respectable lawyer from a respectable Jewish background. He had been to St Albans Grammar School and had until that year worked at the law firm of Rowley, Ashley & Co.

In 1964, Solomon's wife had met Patricia Thompson at an antenatal clinic in north London. David Thompson, her husband, was a butcher who needed a lawyer, and soon Solomon was helping him with his excursions into the world of wheeler-dealery, buying up stakes in a company here, a company there. In 1975, they formed a company, which was called Hillsdown after Solomon's house, to bring Thompson's interests together.

It was in 1980 that they, together with a young accountant called John Jackson, decided to buy Lockwoods. It was a canning business that had gone into receivership, and they paid £3.5 million for it; they thought it was too good a deal to miss.

It was Lockwoods that made Solomon and Thompson realise that there were riches out there for the taking. They discovered that a company that had been mismanaged out of existence by the people at the top could be a goldmine of untapped talent below. It was exactly the same conclusion that Hanson and Green had come to – but it still took the two by surprise. "I was quite naive," says Solomon. "I was shocked by the chasm between good middle management and the people at the top."

Although Solomon and Thompson were mercifully ignorant of many of the ways of traditional British business, they could see clearly that the industry they understood – food processing – was in need of a major shake-up. It had its bad managers but more importantly, it was fragmented. The shift towards large supermarkets had not been mirrored by any consolidation in the companies that supplied them. Solomon and Thompson realised that by bringing together food producing companies, and building up close relationships with the supermarket chains, both sides would benefit. In particular, they thought they could provide the slightly upmarket processed products that were increasingly being sold under supermarket brand names.

In 1982, Hillsdown made its first big purchase, when it paid £39 million for Imperial Foods' poultry division. From then on, as Solomon tells it, there did not seem much point in stopping. "There were so many opportunities," he says. "We could see what we could do, and were amazed that other people hadn't done it before. We did something and it worked, and we did something else and it worked: it didn't seem that difficult." Hillsdown was on a parallel path to the low-tech conglomerates. Its field was quite different, but its techniques were remarkably similar.

# 9

# *More dealing*

Soon after BTR had set the ball rolling with its bid for Thomas Tilling, Victor Watson discovered the purgative power of the takeover attempt. In many ways, Watson was a typical guardian of declining British industry. His grandfather, also Victor, had been foreman at John Waddington, a printing company that specialised in theatre programmes; when the company got into trouble in 1913, Victor persuaded its creditors to let him revive it, and moved it successfully into playing cards and games. In 1936, his son Norman was sent an early version of an American game called Monopoly. He played it, and leapt to the phone to buy up the British rights. By the 1970s, Waddington had diversified into packaging and specialised printing; with its name also on Monopoly, Cluedo and Subbuteo, it felt cosy enough as a constant name in corporate Britain. Victor, who took over as chairman in the late 1970s, was one of Leeds' business leaders: a stalwart of the chamber of commerce and a celebrated after-dinner speaker. He and his two brothers led the publicly-quoted firm with gentle paternalism.

In 1978, they decided to get into the expanding electronics games market by buying a marketing operation called Videomaster from the receiver. The gods disapproved: the main supply factory in the Philippines was hit by a typhoon and covered with four feet of silt. Watson could only watch as other electronic games suppliers undermined his traditional products and in 1980, he had to sell off the Valentines greetings cards business to raise £4 million, more or less what Videomaster had lost so far. When Waddington launched a bomb detonation game weeks after a bomb disposal expert had been killed in Oxford Street, the popular press howled with rage and the game had to be withdrawn. The company, apparently jinxed, slipped deeply into the red.

The losses showed up Waddington's fundamental weaknesses. It had added new businesses without attempting to integrate them properly; an enormous product range was overlaid with different discount systems and massive duplication: it all needed sorting out in a hurry. Watson recognised that he suffered as badly as anyone from being a family boss – he had not the stomach for drastic surgery. So he enrolled David Perry, a six foot six former England rugby captain who had had a successful career in packaging and printing, to do the cutting for him: soon twelve subsidiaries were reduced to seven, sixteen factories to ten and 2,200 workers to 1,760. Perry also shook up the management so that new ideas could be more efficiently exploited: Waddington's became professionally managed.

On May 17, 1983, just as the BTR/Tilling battle reached its height, Watson, who was on a business trip in Scotland, learnt that Norton Opax, a lottery ticket printer a sixth of his company's size, was making an £11 million takeover bid. Richard Hanwell, an accountant and Harvard Business School graduate, had been in charge of the bidder for just over a year and had already shown his predatory credentials with six small purchases and a fivefold increase in the company's share price.

In the nick of time, Watson produced his latest figures, showing that Perry's rationalisation had clawed the group back into profit, and Waddington's shares climbed over the Norton bid: this was not to be Hanwell's entrée into the big time. But at 7.30 on the morning of June 17, just as Watson thought he had scraped past Go, Robert Maxwell rang him and announced: "I am the White Knight, I have come to save you." Watson told him to go away, so Maxwell made a £13 million hostile bid. He wanted to add Waddington to his British Printing and Communications Corporation, already Europe's biggest printer.

Later, he upped the offer to £18 million. Whatever happened, the shareholders could not lose – Waddington's shares, which had started the year at 68p, were now standing at 256p. It was not surprising there was nothing the City liked better than a takeover battle.

On August 25, just as it looked certain that Maxwell would nudge past the 50 per cent mark, Norwich Union, which owned 4.4 per cent of Waddington, changed its mind and decided to oppose the bid.

What had happened demonstrated how completely the fate of British companies was in the hands of a few fund managers. In this case, it was Gavin Mills, investment manager of Norwich Union, who had received a desperate phone call from Watson on the 24th asking for an urgent meeting. Mills agreed, and Watson and Perry drove overnight from Leeds to Norwich: they were waiting on his doorstep when he arrived for work at nine o'clock. Mills, whose Waddington holding was worth only £250,000 out of a total portfolio of £1.5 billion, admitted that he had not thought too hard about the bid, and assumed that Waddington's suffered from all the usual ills of a family-run business. When he met Perry, he realised he had made a mistake, and within an hour the two went away satisfied: Maxwell had been seen off.

But, although the fight had cost Waddington £210,000 and a lot of management time, the bid had done wonders for its share price. As the full fruits of Perry's rationalisation came through in the form of a £3.5 million profit in 1983, the City decided it was something of a glamour stock. Watson and Perry, having been visited with the magic of a 480p share price by the summer of 1984, took full advantage of it by buying two specialist printing companies and making a £5.2 million rights issue.

The trouble with rights issues is that they can upset the shareholders who are being asked for money. As Maxwell still owned a quarter of the company, and believed that the share price was high only because the City was expecting him to bid again, he was furious, and unsuccessfully demanded that the issue be cancelled.

He appeared not to take up the shares under the rights issue – so defusing bid speculation – but in fact his allocation found its way back to him and his allies. In October 1984 he offered £44 million for Waddington. The City clapped its hands with glee, and pushed the share price up to 525p. This time, Maxwell made no pretence of geniality. During the first battle, he had phoned Watson fourteen times; now, war was waged incommunicado. Waddington's merchant bank, Kleinwort Benson, unimpressed with what they saw as Maxwell's trickery over the share issue, took its gloves off and made much of the mysterious real ownership of his Liechtenstein-based holding company. In the end, it was Maxwell's own oddball bid tactics – no possibility of an extension or increase of the offer – that defeated him. Just as it looked as though the Waddington share price would slip below the 500p offer price, record profit figures came storming over the horizon; on the critical day, the shares were strong enough to beat the bid. Waddington kept its independence, but it was a very different creature from the one that had blundered into the Philippines seven years earlier.

Riding the bull market, the new-style predators were busy consolidating their mini-empires, in the process mopping up those who had not the luck or good judgment of Victor Watson.

On Boxing Day, 1984, Nigel Rudd was lying in his bath reading the latest *Investor's Chronicle*, when he noticed an item about a company called J. and H. B. Jackson. It was, he read, a Midlands engineering business with interests in forging, engineering, and a Ford main dealership. It had grown steadily through the 1970s, and had survived the recession well, although in 1984 its profits had dipped. What really interested Rudd was that it had £20 million of assets, of which £10 million or so were tied up in gilts and shares: the chairman was, it seemed, keen on dabbling on the stock exchange. It had a modest price/earnings ratio. Rudd rushed to the telephone to tell McGowan that he had thought he had found the key to Williams' next move.

Ley's had been turned round with classic Williams techniques: it was pulled out of mass production, prices were put up and 660 jobs were cut. Since then, the group had bought a handful of new car dealers and Garford-Lilley, a plastics company. It had just made its first, modest, profit but consisted now of an unremarkable collection of businesses with £10 million of assets and an uncomfortable £11 million of borrowing. Rudd reckoned Jackson's could put Williams on a sound base: the gilts and shares would, after all, immediately wipe out most of its debt.

Working closely with David Challen, an advertising man turned Schroder's merchant banker, Rudd and McGowan offered shares worth £24 million for Jackson: twice the total they had spent so far on acquisitions but, thanks to a share price that had multiplied tenfold since 1981, quite manageable. They had already talked to Jackson's chairman, Philip White, and he, while publicly dismissive, had made clear that everything had its price. When the bid was made, the City cheered and pushed Williams' share price up to the point where the offer became irresistible to Jackson's shareholders; on May 1, Williams took delivery of the company. Rudd's bath-time thoughts had worked out beautifully: debt gearing was reduced from 100 per cent to 33 per cent. Had the bull market stalled before this deal, he knew, Williams would have been destined to stay small: now it had strength to grow mightily, come what might.

Other potential Lord Hansons kept popping up, although four in particular had been marked out by the City for future greatness – Williams, Evered, Tomkins and Suter: the WETS. Evered had made some modest purchases but, in the summer of 1985, seemed to be on the point of a bold move. The Abdullah brothers had gathered a group of Saudi investors together, and had bought 11.6 per cent of TI Group, an engineering giant that was stuck in the doldrums. With Evered's market capitalisation half that of TI's, a full bid did not seem unlikely. Suter's David Abell finally got into gear at the end of 1984: having slowly built the company up with a handful of small acquisitions, Abell finally took on two engineering companies, Francis Industries and Lake & Elliott. But he was also adept at buying and selling chunks of other companies. He, more than the other WETS, was a man of finance.

Tomkins joined the brigade of predators later than the rest. It was interesting for the impeccably analytical style of its new chief executive, Greg Hutchings. Armed with an MBA and a spell as an acquisitions spotter for Lord Hanson, he decided to look for a company of his own. His technique was straight out of a textbook: he took a list of 350 companies valued at under £10 million, and narrowed them down to a shortlist of ten by sitting down with a pile of balance sheets and a calculator. He sought out companies that were basically sound, but had an undervalued share price, and this way, he came up with FH Tomkins, a Midlands buckles and fastener company, well-run but little rated by the City. Using his Hanson credentials, but refusing to disclose the name of his intended target, he persuaded City institutions to back him with £2 million and, on January 24, 1984, took it over and became its chief executive. With classic Hanson techniques – tight financial controls, bonus and share schemes for managers, and a demand that they aim for a 20 per cent return on capital – Tomkins started to roll. His presence alone had caused the share price to double: when the City saw what he was doing, it multiplied, and he started to go shopping. By the summer of 1985, he had bought Ferraris, a piston distributor, Hayters, the lawnmower maker, and six GKN subsidiaries.

The WETS were distinctive creatures of the bull market, but not the only ones. They were conglomerates, characterised by a willingness to involve themselves in a wide range of industries. They had cousins, specialist predators, who would choose a sector, and start buying. They included Siebe, founded by an Austrian officer who came to England and developed the first diving equipment but which was now being turned by Barrie Stephens into a heating and ventilation giant; BBA, maker of conveyor belts and brake linings, which was transformed by John White with £150 million worth of acquisitions and disposals between 1983 and 1985; and Meggitt, which Ken Coates was building into a "mid-tech" group, specialising in aerospace, defence and electronic components.

A branch of the specialist predator family included Hillsdown and Coloroll, companies that were taking complete industries, and merging them into shape.

Hillsdown was the "nice predator". Harry Solomon and David Thompson never made hostile bids. Instead, they identified companies that they thought would fit into the Hillsdown empire, and tried to persuade the owner of the benefits: their targets were invariably private companies.

In fact, Hillsdown itself only became public in February 1985. Although it had already picked up a reputation for making acquisitions sparkle, it took much persuasion by Solomon and finance director John Jackson to convince the City that there really was something special about the Hillsdown treatment. Not only did results improve, people seemed to be happier, more motivated. As Hillsdown hoovered up thirteen companies by the end of 1985 – the biggest costing only £6.5 million – the City decided this was a company to back: the share price, inevitably, rocketed.

Coloroll was floated in April 1985: the launch was disastrously received, but John Ashcroft cultivated City fund managers so assiduously that they soon learned to love him. He was clever, immodest and highly ambitious, and under him Coloroll had a great effect on its industry, even though much of the groundwork had been laid before he arrived. But Ashcroft was an interesting fellow for, with his obsessively analytical mind, he was almost a caricature of the new professional manager, albeit with a marketing rather than financial background. He was born in 1948 in Wigan, the son of a quarry worker, and had gone via grammar school to the London School of Economics, where he was known mainly for his athletic prowess. After management training with Tube Investments, he moved into the international marketing side of Crown Wallcoverings, part of Reed. His colleagues noticed his unusual turn of mind when he produced a complex report which, by analysing inflation and exchange rates, could predict which export markets were likely to be profitable over five years. Many assumed he had been to business school – in fact, apart from a short course at the London Business School, his approach was self-taught. According to him, he had "a jigsaw mentality – before I could understand anything, I had to have a much bigger picture, to see how it fitted together".

Ashcroft joined Coloroll in 1978. By then, the group was already making a name for itself under the guidance of John Bray, a marketing man who had been brought in by the controlling Gatward family in 1971. He had imported a top design team and launched Dolly Mixtures, a range of inexpensive coordinated wallpaper and fabrics in the Laura Ashley style, which, reinforced by a uniquely supportive service to retailers, soon started to cut a slice out of a shrinking market. Coloroll's sales, which had been £1 million in 1970, reached £11 million in 1979.

Ashcroft, who became managing director in 1979 while still only 30, reinforced the marketing-led ethic, and started applying his analytical techniques to the company's operations. He carried out "a poor man's McKinsey" on Coloroll, looking at every facet of the business. He replaced bi-monthly accounts with weekly profit and loss accounts and with monthly reports that had to be out within five days. He reconfigured the machinery layout – before, a half million pound piece of machinery was working short time for lack of another machine that cost £15,000. He put in a new sales order system, new computers, new administrative systems.

Then he put together his "white book", an assessment of where the business was, where it was going, and where the market opportunities were. Next came the "yellow book", a detailed marketing analysis of the entire business; it showed where every roll of wallpaper was coming from and which retail group bought it. It was used to evaluate which rival businesses were most vulnerable, and which accounts should be targeted.

As a result of all this analysis, Coloroll continued to carve out new segments of the market, creating new designs for each type of customer. It tapped into the newly emerging DIY chains, B & Q and Texas, and set up special sales teams and rapid delivery programmes to service them.

The 1980 recession caused massive destocking, and Coloroll laid off 10 per cent of the workforce. But within six months, it was back to full strength; profits were dampened only slightly and Mrs Thatcher singled the company out as a role model in her speech at the opening of Parliament that year. In early 1982, when Ashcroft took over from Bray as chief executive, the Gatward family sold out to a group of institutions in a semi-flotation. The group moved abroad; after a disastrous attempt to launch Dolly Mixtures in the US, it bought into a large American wallpaper group, Wallmates. By the time it finally reached the stock market, Coloroll's annual sales were running at almost £60 million, with profits of about £5 million. Ashcroft, supremely self-confident, had little difficulty in selling himself to the City, and could now start building a real empire: he wanted to create "a fully integrated supplier" of things for the home. To that end, he started buying other companies, including Staffordshire Potteries (to make mugs), Fogarty (the duvet company) and Wallbridge, a carpet manufacturer.

Meanwhile, managers who had been fighting in the trenches were wondering if perhaps the acquisitions game might not be more agreeable, and sensible. In August 1982, Tim Hearley's Coventry Hood, now called CH Industrials, had a market capitalisation of £2 million. After the failure to rescue MG, it found itself with a half share in Aston Martin, and Hearley – who had bought out his mother-in-law and was now the major shareholder – set about trying to make the most of his rather unusual position. Apart from the share in Aston, the group was in a parlous state: its hood operation closed, its activities were confined to trimming work and the barely profitable chemical subsidiary, Beaver. Hearley kept the company ticking over by developing thirty acres of land in Buckingham, which had come with Beaver in 1977.

With a new managing director, an accountant called John Kinder, and the small amount of cash coming from Buckingham, he decided to go for growth. He reckoned that even if hoods were out of fashion, there must soon be a resurgence of open air motoring and that car companies would soon start to fit sunroofs as standard equipment. So he bought Tudor Webasto, a poorly sunroof maker, and formed a joint venture with Webasto of Germany to sell its range to manuufacturers. It was a shrewd move: soon, car companies were indeed fitting sunroofs, and Tudor Webasto came to dominate the market.

Meanwhile, Hearley was having fun and games with Aston Martin. Victor Gauntlet, who had made a fortune with Pace Petroleum, took over the other half of the company, and Hearley found himself joint chairman. Aston represented a quite different side of British industry: its handbeaten, fabulously expensive cars were what Britain was still best at: they went with cashmere sweaters and Wedgwood pottery as products that could never be bettered by foreigners, simply because they had to be British. Yet, like Wedgwood, Aston Martin was in trouble: indeed, it had rarely been out of it. It suffered from a craft version of militant unionism: twelve panel beaters would regularly stop work to demand more money or better conditions, and each time the management gave way. The company had been through any number of owners since the war, and the panel beaters had learnt that if one proprietor retired, financially exhausted, another prestige-seeking millionaire would always pop up.

Hearley's advantage was that he was not after prestige (he was not even particularly interested in cars) and he tried to introduce professional management techniques. But the panel beaters beat him all the same. Just when he thought he had outmanoeuvred them by switching to an outside firm, they found out and picketed it. In 1984, by which time the other half of the company had been passed from Gauntlet to a group of Greek Americans, he sold out. But he took a bit of Aston Martin with him: its main strength, he decided, was its team of thirty development engineers, who were sitting around with little to do. He put them into a new company, jointly owned with Gauntlet, and called it Aston Martin Tickford. The idea was that it should design and even produce "specials" that were not worth the manufacturers' while making. Almost immediately, Tickford won the contract for the cabriolet version of the Jaguar XJS. City analysts started to notice that Hearley and Kinder were an astute pair – CHI wasn't making much money, but it was digging itself niches in potentially profitable areas. Its shares started to move up, and Hearley looked around for other companies to buy.

On January 25, 1984, Brian Taylor announced first that his company was now called Wardle Storeys, and second that it had made a profit of just over £1 million in its latest financial year. The Storeys takeover had been completed at a knockdown price – £4 million, of which £3 million was redundancy costs. In six months, the Lancaster factory had been closed, product lines had been moved around between the three remaining factories, and four sales forces had been merged into two. It was all pure management carried out with ruthless efficiency – that was what Taylor was best at. He was still making the same sorts of products as before – mostly converting PVC into plastic sheeting – but was big enough, and efficient enough, to take on international competition. Wardle was moving into niches, where competition was limited and profit was higher. As well as Dedpan, now softening the clunk in car doors around the world, Wardle's products included plastic for prams and awnings for Harrods.

In November 1984, Taylor received his reward for years of slogging: Wardle Storeys was floated on the stock exchange, and he picked up £3 million. But he still had £5 million worth of shares, and made no secret of the fact that he was not stopping there. The City licked its lips: it had taken a shine to this industrial Clint Eastwood, and had marked him down as a future Lord Hanson.

\*

## More Dealing

Glynwed International was, by West Midlands engineering standards, a very well run firm. Both Lesley Fletcher, who was chairman through the 1970s, and Gareth Davies, who took over in 1984, were accountants, but they let the subsidiaries (mostly run by engineers) alone as long as they performed. Head office provided only services that the subsidiaries could not provide: in the late 1970s, for example, planning director Nick Boucher was able to tell divisional bosses that Britain was about to plunge into recession. Glynwed's was a good way of combining the skills of bean counters and spanner brains.

But Davies recognised that in a bull market, being good at what you do was not enough. Once upon a time, in the late 1960s, fast-growing Glynwed had been a glamour company in the City. Not any more – now it languished with the rest of the engineering sector and did little to let the City know what it was up to; that, Davies knew, was not sensible at a time when the stock market was roaring and predators were on the prowl. So, borrowing some of their techniques, he set about getting his share price up. He commissioned a survey of City folk and found that they thought Glynwed was a typical Midlands metal basher, with borrowings too high and an overdependence on South Africa (which had, by dint of politics and a falling gold price, turned from a lifebuoy into a millstone).

Davies understood that what City people like is a "story": with Williams, it was the young accountants taking on a decrepit Welsh foundry; GKN had sold itself on the strength of its constant velocity joint. Glynwed had to provide a similar bit of titillation. Even though he did not consider that Glynwed had too much debt, the City did, so he decided to cut it. He also offloaded the South African operations and started a programme of educating analysts in the complexities of the steel and engineering market. But the climax of his story was a bold one: he told the world that Glynwed would increase its earnings per share by 20 per cent a year, while maintaining a return on capital of at least 25 per cent and keeping its debt/equity ratio below 40 per cent. These were astonishingly ambitious targets – the sort of figures that the most aggressive predators aimed for – and the City was duly astonished. So were the managers of Glynwed companies, who now had to run very fast indeed. It did not matter how they met the targets. Davies told them – they could reduce costs, introduce new products, find new markets, anything – but they must meet them.

Unhappily, not all of British industry was on the turn. Out in the trenches, the recession was claiming its last victims.

# 10

# *Death of a giant*

On September 3, 1984, Acrow went into receivership. It was a traditional large British engineering firm, built up by an engineer mogul with all the strengths and vices of Austin, Morris or Herbert. But there was a twist in the story, for William de Vigier was Swiss. He had come to England as a young engineer in 1936, and set up his scaffolding system company on the back of £150 saved while working in Switzerland.

He adopted all the faults of the British paternalist as though he had been born to them: a brilliant salesman and showman, he was incapable of delegating. De Vigier briefly tolerated a managing director in the mid-1970s, but disposed of him when he suggested raising money through a rights issue: that would have threatened his personal control of the company.

The error that led to Acrow's doom was its purchase of Coles Cranes in 1972. Immediately, it found itself in head-on collision with the Japanese as they swung into the market. In 1980, the UK recession combined with Japanese penetration to bring Coles to its knees; when the Middle East construction market crumbled in 1982, it came close to expiring.

De Vigier tried to market his way out of trouble, slashing prices and mounting a massive international sales drive. Profits rose slightly in the year to March 1981, before sliding ever downwards. With pressure from the banks mounting, he called in a tough Scot called Norman Cunningham to mount a rescue. Cunningham chopped 2,500 of the 6,000 workers, worked out a restructuring plan with the banks, and managed to bring Coles to break even – ironically, it was then the only British crane maker to be covering its costs. But by early 1983, Acrow's losses were mounting and its debts had reached £50 million. Cunningham came within a tad of selling Coles off but, as the deal fell through, it became clear the group would not be able to pay its interest charges for 1984. Receiver Michael Jordan moved in and took a full page advertisement in the *Financial Times* advertising his wares: the limbs of the Acrow Group.

Acrow was the last of a comparatively small group of companies to be pushed by recession into receivership: it included Capper Neill, Stone-Platt and a boxful of toymakers. These were the companies not even the Bank of England could save. But, as the bull market roared, there were plenty of other great names that disappeared – not into oblivion, but into the jaws of the predators. Of these, none was greater than Dunlop.

The story of Dunlop's decline combines all the elements of 1980s industrial turmoil: bad management, bad strategy, City predators, mysterious stakeholders – and the Japanese. It shows, more than anything, just what a complicated business business can be.

In 1888, the Belfast vet John Boyd Dunlop fitted a pneumatic tyre to his son's tricycle; he patented his invention and laid the foundation of the world's first tyre giant. By 1934, when the *Financial Times* thirty share index was created, Dunlop was part of it: it had become one of the mighty forces in industry, and nothing symbolised its strength as much as Fort Dunlop, the 300-acre complex outside Birmingham that even had its own fire station and hospital.

By the 1960s, Dunlop employed 100,000 people around the world and was expanding rapidly in Europe under the chairmanship of Reay Geddes: his family, along with the Beharrells, had been running the group for many years. At the end of the decade, Geddes decided that the company needed to get even bigger – tyre technology was changing rapidly and would need vast capital expenditure – so he instructed his new director corporate planning, the former financial journalist Campbell Fraser, to find a firm that Dunlop could either buy or merge with. Fraser suggested the cable company BICC: Geddes at first accepted the idea, then changed his mind. Next possibility was the US Rubber Company (later Uniroyal), but that foundered on anti-trust problems in the US.

Suddenly, Geddes announced he had found the ideal partner. It would be Pirelli, the Italian-Swiss tyre and cable maker. "They are our class of people; our style of management," he told his directors.

The geographical fit was good: Pirelli was strong in Latin America and southern Europe, Dunlop was big in the Commonwealth and North America.

Geddes brushed aside worries about the opacity of Italian and Swiss accounting systems, and proposed that the merger would be a sort of marriage, with each company taking a large minority stake in the other. In October 1970, Leopoldo Pirelli arrived in London and announced that a proposed change in Italian tax law meant that the merger must be pushed through before the end of the year. And so, in a great rush, it was.

At that stage Pirelli was forecasting a £3 million profit for its Italian tyre operation in 1971. But in mid-January, when the coordinating central committee of the new venture held its first meeting, Leopoldo Pirelli casually produced a piece of paper showing that he was now expecting a £7 million loss. In the event, it was £18.3 million. Dunlop's managers were horrified, but could do nothing – the agreement had no escape clause.

Pirelli's accounts had lived up to the Italian image. Dunlop managers, backed up by Price Waterhouse, had gone through all its figures – as they thought – for several years past. But they had failed to spot that while the group was technically good, it suffered from British-scale overmanning and union problems.

At the time, the union was seen as a great step forward: the first pan-European merger, the shape of things to come. "I went round business schools explaining how we had put it together," Roy Marsh, a Dunlop director, recalls. But there were fundamental problems with melding the two companies. Although linked through committees at operating levels, financial control was always kept separate. "If Italy hadn't gone wrong, it would have worked," Marsh believes. "With three or four good years, we could have gone on to a unified system of management." As it was, there were deep divisions caused by a different approach to problems. The Pirelli family was prepared to go on investing in the loss-making operations; the more cautious British were not.

Dunlop and Pirelli, in common with other manufacturers, had been switching over to steel-belted radial tyres, which lasted for 40,000 miles instead of the 15,000 or so that could be expected from traditional cross plies. Inevitably, demand for new tyres was slashed and in the 1970s, as the oil crisis braked the car market, overproduction built up. Only Michelin of France, which had invented steel-belted radials and which could make them most efficiently, was equipped to withstand the mounting pressures. Dunlop, which had only one reasonably efficient factory, at Washington in the North-East of England, was not.

Nor was its management able to adapt. With an oppressively hierarchical structure, it found the greatest difficulty in making changes. Dunlop was technically advanced, but suffered hopelessly from lack of coordination. On the one hand, its one great technical stride of the 1970s, the Denovo tyre, was pushed on to the market well before it was ready, and was a disaster; on the other, the Birmingham Tyre Technical Centre, which was one of the leaders of tyre technology in the world, would develop products that no one had any use for. It was a classically arthritic British company.

Until reality hit home, the board was feeling that it was doing really rather well, despite the Pirelli albatross. Dunlop had its best results ever in 1976, mainly because it was making progress with consumer goods, including tennis rackets and shoes, and in specialist rubber products. In some areas, such as aircraft brakes, it was carving out useful technologically advanced niches.

The shock came towards the end of the 1970s, when the UK tyre operations went into loss. The Labour government wanted the NEB to take a stake: Dunlop turned the offer down and started cutting vigorously. The Speke factory on Merseyside, a mile away from the doomed Triumph plant, was closed in 1979. That was just the start. As the recession poured in, losses mounted and Sir Keith Joseph was forced to fish £6 million out of the government wallet in the summer of 1980; it would be used to modernise the ageing plant at Fort Dunlop and was the biggest subsidy to a private company the new government had made.

By now, the tables had turned: Pirelli was having to support Dunlop, and both sides agreed that the knot must be untied. It proved remarkably simple, although Dunlop's delicate finances were buffeted by the £40 million it had to chop from its reserves as a result. By 1982, when the union was finally dissolved, it had cost Dunlop about £70 million.

Fraser and his new managing director, ex-Treasury mandarin Alan Lord, had little time to ponder the lessons of history. They were trying to cope with the £7 million the group had lost that year. After provisions for redundancy payments and the like had been included, the total loss came to £83 million.

Fraser got on the wrong side of practically everyone when the annual report for 1982 showed that he had had a 21 per cent pay rise, to £82,000. As chairman of the CBI, he had repeatedly preached the virtues of pay restraint, most famously saying that Britain had gone "bonkers" over pay. For a man who had joined Dunlop as head of public relations, the rise was an extraordinarily inept move; many thought it cast him as an example of the worst breed of detached British manager. The next year, 200 senior managers got bonuses averaging 20 per cent. Said one: "We thought it was to keep us quiet about the chairman's rise."

Ian Sloss, a down-to-earth Scot, arrived at Fort Dunlop as personnel director in 1980. He had just finished closing down Speke, where he had been works manager. He found the morale at Fort Dunlop dreadful: the local managers had abrogated responsibility to the unions, and felt that head office in London had little idea of their troubles. It was easy enough for the men at Dunlop House in St James's to forget the practical problems of making tyres: they had their own problems. There were attempts to improve productivity – a team from the American consultancy McKinsey had just finished a study when Sloss arrived – but everyone could see that the whole machine was going one way: backwards. The recession had gripped the motor industry, and Sloss could do little but cut and cut again. His token efforts to introduce quality circles and close seven executive dining rooms earned few thanks: "You are confusing industrial democracy with egalitarianism," deprived executives told him coldly.

Its wide geographical spread saw Dunlop through the 1980 recession, and in 1981 it seemed on the point of recovery. Then came the world slump in 1982 (the French plants were hit particularly badly) and the group headed off down the slippery slope. In the six years to 1983, Dunlop's British workforce had been cut from 48,000 to 25,000, with the tyre workforce falling from 11,500 to 3,500, but capacity remained stubbornly ahead of tumbling demand.

Although 60 per cent of Dunlop's business was tyres, the other divisions, which ranged from vast rubber plantations through industrial belting to tennis rackets, were being starved of investment as their profits were diverted to keep the tyre division alive. The non-tyre managers became increasingly restive.

Fraser and Lord had decided in 1981 that Dunlop must get out of tyres in Europe. It was a brave decision, cutting the original heart out of the company. But who would buy a loss-making, not very efficient tyre maker in an oversupplied market?

In 1908, John Boyd Dunlop had founded Dunlop Japan in Tokyo. In due course, it became Sumitomo Rubber Industries, and Dunlop slowly ran down its shareholding as the rapidly-growing company demanded more and more funds. By the early 1980s, the holding was 40 per cent. But the Japanese government objected to even that level of foreign ownership; it was proposing a law that would make

Sumitomo lose its Tokyo stock market quotation unless Dunlop sold out. Lord saw a god-given opportunity, and grasped it with both hands. Sumitomo could have its shareholding back, he told the Japanese management, but only if it also agreed to take over Dunlop's European tyre operations. "They coughed and spluttered and said we have no money," says Marsh. But Lord had them over a barrel and, pointing to other Japanese companies that were forming links with European and American manufacturers, coaxed them to the negotiating table.

No Japanese company had ever bought a major going concern in Britain: investments had always been in nice new factories well away from unions. Sumitomo did not want Fort Dunlop, with its seven unions, but was prepared to take the relatively new factory at Washington, Tyne and Wear, as well as the profitable German plants. Lord would not budge, and forced one concession after another on to the Japanese. By the time a deal was announced, in September 1983, they agreed to take the whole shooting match except the French factories, and would pay £112 million for it. Almost immediately, the French operation filed for bankruptcy, and Sumitomo ended up buying that, too. For a super-cautious Japanese firm with little overseas experience, it was a courageous move.

Meanwhile, a strange sub-plot had been developing. A smallish Malaysian investment group, Pegi Malaysia – led by Abdul Ghafar Bin Baba, vice chairman of the ruling party – had started buying Dunlop shares in 1978; by April 1983 it had built up a 26 per cent holding in Dunlop, and was given two seats on the board. The Dunlop management, while cooing contentedly in public, had a low opinion of the Malaysians: "unreliable people", they thought, and they were unsure what they wanted. Would they make a bid for the group, or not? Did they even want to? Dunlop thought the holding was probably a device to force it to sell its Malaysian assets, so it tried to shake Pegi off by selling its vast rubber plantations to Chinese Malaysian interests. This was a time when there was great tension between Chinese and Malay communities, and also between the Malaysian and British governments: the prime minister Dr Mahathir had launched his "Buy British last" campaign after a row over airline rights. Was Pegi backed by the Malaysian government in an attempt to teach the British a lesson? If not, where did the £40 million it had spent on shares come from? The Sultan of Brunei perhaps? Dunlop managers spent much time trying to find out, and failed. To Marsh, then director of corporate affairs, it was "a total bloody distraction".

At the end of 1983, Sir Campbell Fraser stepped down as chairman as gracefully as he could: the Malaysian directors played a part in his removal. His £137,000 payoff, cheap Rolls-Royce, title of president and Mayfair office sent ripples of outraged mirth through the City, which reportedly pinned most of the blame for Dunlop's failure on his shoulders.

Sir Maurice Hodgson, who had just handed ICI over to John Harvey-Jones, took over as caretaker chairman. He found a group with £1 billion turnover, 27,000 employees, some impressive technology, especially in aircraft brakes and best of all, no tyre business. But Dunlop had just been dropped from the *Financial Times* thirty-share index and was running out of money fast: it would have to wait many months before anything came in from Japan. Its shareholders' funds – the funds put up by shareholders plus its reserves – were getting dangerously depleted, and with them the group's ability to borrow. The borrowing limit was one and a half times shareholders' funds, and that was about to be breached by debts that were already standing at £450 million. Sir Maurice had to ask for the limit to be raised to £600 million: the banks, in agreeing, were now in a position to dictate every move at Dunlop.

Immediately, schemes popped up to rescue the group. One came from former joint managing director John Simon, who had resigned in protest over the Pirelli deal in 1972. He was trying to put together a financial consortium to bid for Dunlop. His proposal came to nothing, but was ahead of its time: it would now be called a management buy-in.

Even though running losses were staunched in 1983 – a £17 million pre-tax profit was recorded as the market started to perk up – Hodgson decided to load all the bad news he could into that year's figures to give himself a clean sheet. Likely redundancy payments and the costs of closures were added, bringing the total loss to a terrifying £166 million. Shareholders' funds were now down to £100 million, a fraction of what was owed to banks; Dunlop was on the verge of bankruptcy.

But there was still hope. By the middle of 1984, optimists in the financial press were saying that Dunlop stood a chance: Airbus Industrie had ordered Dunlop carbon brakes for its new aircraft; Austin Rover had put in a three-year order for Dunlop steel wheels. What was needed was a "financial restructuring"; like Mexico, the group needed an extra injection of cash to give it a chance of paying off the money it already owed. The forty banks involved agreed to defer the £200 million they were owed in 1984 while a reconstruction plan was worked out.

Through the summer of 1984, the banks laboured on Project Robin (they always use bird names for lame ducks), which would put the group back on its feet by selling assets, transforming existing debt into some sort of equity, and raising new money. Although Barclays and National Westminster were fronting the group, the Bank of England was intimately involved. The aim was to create a company that could start again: Hodgson called it Son of Dunlop.

It was clear that the banks wanted to replace Alan Lord, now chief executive. Although he had masterminded the Sumitomo sale, and had pushed through some of the nastiest cuts, he was criticised for lacking a long-term strategy and for arrogance: the former mandarin had an ability both to cow his subordinates and to irritate those he needed to impress.

Newspapers buzzed with rumours that Sir Michael Edwardes, who had just left ICL after it had been taken over by STC, was the front runner. His history of troubleshooting on a grand scale made him an obvious choice. But Hodgson did not want him, because he thought they could not work together. He was probably right: Hodgson, astute and careful, had been brought up in the step-by-step cooperative world of ICI; Edwardes prided himself on being an individualist action man. Hodgson announced his chosen candidate, inevitably an outsider and this time an American called Eugene Anderson. He was vice president of the textile giant Celanese.

Anderson was all set to get on a plane at the beginning of November when he was told to unpack his bags. The banks had decided that Edwardes was to be their man after all; on November 8 he installed himself in Dunlop House. The day before, Hodgson had resigned and Lord and the three senior directors (including Marsh) had been fired. Edwardes offered to meet them: they told him to get knotted. They did, at least, have three years' salary to keep them going, golden handshakes totalled £880,000. The press, gleefully dubbing them the Gang of Four, clearly saw the purge as the end of the "old management". Edwardes filled their places with bright young men from ICL.

He immediately chopped the group into seven operating centres, in his approved decentralising style, and set about working out the details of the financial restructuring. Objections from Pegi slowed the process down, but on January 11, 1985, the arrangements were announced. They involved converting bank debt into equity and issuing new shares, both to existing holders and to new investors. Dunlop House and the National Tyre Service would be sold off, as would the Malaysian and New Zealand offshoots. The original shareholders would be left with less than 14 per cent of the new company.

This was the first time that British banks had agreed to take a major holding – 10 per cent – in an industrial company. It was a device that was normal in Germany, and now the Bank of England decided it was necessary here. It, and institutional investors, had become fed up with the willingness of clearing banks to put companies into receivership when there was still a chance of them being revived. The most notorious case was Stone-Platt: in 1982, a year after institutions had put new money into this textile machinery maker, Midland Bank had called in the receiver. With Dunlop, the clearing banks could call all the shots, because they were owed so much, so they had to be doubly discouraged from doing the same again.

Edwardes was also a great believer in providing executive motivation. As part of the deal, he would be able to buy 21 million shares in Dunlop at 14p each from October 1987 on: this "share option" meant that he would make £2 million if the share price made it to 25p. Two other directors had lesser "equity kickers" – between them, the three stood to own 5 per cent of Dunlop.

When the shares were floated on January 18, they shot up to 31p: Edwardes had made a £3.5 million paper profit. By the time Owen Green (now knighted) walked into his office at 9am, he already knew why. BTR was offering £33 million for the whole of Dunlop: it had taken advantage of the complexity of the restructuring package to snap up 25 per cent of the group's preference shares from the institutions. Even though that gave Green only 5 per cent of the total equity, he had put an armlock on the restructuring: 75 per cent of preference shareholders had to approve it. It was little short of blackmail, but the financial press was delighted with the ruse: they quickly decided that Dunlop and BTR would fit together excellently. Green did not really care. As an apostle of the conglomerate, he considered there was nothing that could not benefit from BTR's management style.

The Dunlop old-timers were not wholly surprised. In the 1960s world of rubber, BTR had been the poor cousin of mighty Dunlop. Green had been sniffing around for some years, with the idea of picking up some of its divisions. But the audacity of the move took the City's breath away. Usually, when there is a bid, the share price of the target goes up, and that of the predator goes down. This time both went up, a sure sign that the City thought the whole thing was a terribly good idea.

Edwardes' share options enabled Green to cast him as the baddy. By calling the 23p bid offer "grossly inadequate", Sir Michael was implying that he was entitled to a lot more than £2 million on his options. Sir Owen claimed that he would not have made the bid had Hodgson stayed in charge. "But everything changed when he took over. He is backed by God knows whom at his price of 21 million." The predator was casting himself, rather implausibly, as a knight who would rescue poor old Dunlop from the avaricious South African.

Edwardes retorted that BTR was being opportunistic. It was: opportunism was, to Green, a prime virtue. The City sat back for a lucrative battle between two of the very best new-look businessmen. Both were characters. Anthony Sampson describes Edwardes: "He looks, as he is, like a being from another world: a tiny, agile man with a small head and narrow eyes, who moves like a tree-creature which might drop on your shoulder."[1] He was a man who worked best on the defensive, where fleetfootedness and aggression counted for most. Green was altogether cosier. Looking, talking even, like a somewhat filled-out Michael Hordern, a wisp of hair flopping untidily over a balding pate, he was a slightly eccentric but kindly uncle: fierce as blazes when he wanted; impishly charming when he did not. He had never been on the defensive in his life; his career had been nothing but gobble, gobble, gobble. Both men, needless to say, had razor-sharp minds; and both were "outsiders".

That it was going to be a tough match fight became clear when Dunlop's share price shot to 36p, against the 23p bid offer. Almost immediately, Edwardes gave up his share options.

If "City opinion" was all behind Green, the clearing banks – who had all the power – were not. They were affronted by his buccaneering tactics and rejected his blandishments. And, when the small shareholders (who held an unusually large share of the equity) announced that the BTR bid "approached piracy", Sir Owen found himself a couple of games down.

From the sidelines, an all-1980s phenomenon suddenly popped up – the Wall Street arbitrageur Ivan Boesky had apparently started buying Dunlop stock. "Arbs" were speculators who would buy shares during takeovers, and make killings as they were pushed up by the bidding. Boesky, their "king", was reckoned to have made more than $250 million on his dealings. There was a flutter of excitement, before Boesky made it clear that this game was not for him.

The two wizards battled on. As Sir Owen claimed that Edwardes' refusal of the bid was "illogical, incomprehensible and indefensible", and Edwardes claimed that the bid was "ludicrous", shareholders became more and more confused. BTR advertisements were curtailed after the takeover panel declared them misleading: in the absence of hard figures, all either side could do was to throw rhetoric at each other. At one stage, BTR wrote to Dunlop's merchant bank, Warburg's, saying that Dunlop's directors could find themselves personally responsible for the postage if they sent letters to BTR's shareholders. Meanwhile, Warburg's was busy trying to restructure the restructuring so that BTR could no longer block it. The clearing banks went along with the scheme. The first few games were going to Sir Michael.

Sir Owen managed to mollify the clearers: they might, they said, agree to continue financing Dunlop if BTR bought it. But the real question was over the share price. It was pushed up to 46p, double BTR's offer, and rumours about a counter bid started to circulate.

Towards the end of February, the re-restructuring was announced: it would enable all the money to be raised via a rights issue, with the banks agreeing to pick up the shares existing shareholders did not want. Preference shareholders would be irrelevant. Sir Owen was at least two sets down. The only way he could buy Dunlop was by making an acceptable offer. But what was acceptable? No figures had been published for 1984, and it was impossible to tell whether the group had really pulled out of its financial nosedive. One stockbroker said the shares were worth 52p; another said they were worth nothing at all.

Edwardes took another quick game when he announced that he was close to selling Dunlop's profitable American tyre operation to its management for $120 million. The share price climbed to 51p. Green would have to do something spectacular to win the match.

He did. After lunch on March 8, he walked into Edwardes' office again. This time, he stayed for three hours and at 6pm, signed a deal agreeing to pay £101 million for Dunlop – 63p a share. Sir Michael had revealed figures that showed that Dunlop had halved its losses in 1984 and that, under the rationalisation plans he had set in motion, debt should be down to £100 million by the end of 1985; crucially, shareholders' funds still stood at £58 million, much more than anyone thought. On March 27, BTR announced it had acceptance from more than 50 per cent of Dunlop's shareholders. Dunlop was no longer a company; now it was just a trade name.

Pegi found itself with a stake worth £24 million; it had spent twice that on building up its shareholding. Its directors admitted the whole adventure had been a terrible mistake. Sir Michael Edwardes went off at the end of April, in search of another company to rescue.

# Part Four

## *Time for doubt*

11

# *Better weather*

By 1984, all the action in industry seemed to be centred on two fronts. One was the City of London, where the stock market's bull run led to one takeover battle after another. The other was the coal fields, where the government was squaring up for fifteen rounds with Arthur Scargill; the miner's strike erupted in May and ended with defeat for Scargill the next March. It damaged a few firms, notably those that relied on the NCB for supplies, like British Steel (which reckoned it lost £180 million as a result of the strike) and suppliers, with whom the coal board normally spent £1.3 billion a year. Most managers, though, welcomed the strike as the final sealing of the militants' coffin.

The economy continued to improve. The main motor was the increase in domestic spending, much of it financed by credit. As inflation and interest rates stayed low, and house prices, at least in the South-East, started to lift off their four-year plateau, those consumers who still had jobs were more inclined to borrow. Much as the government exhorted companies to limit pay, wage increases consistently kept ahead of inflation, biting into the benefit of increased productivity, but also giving the economy an extra consumer boost.

By the summer of 1984, vacancy signs were appearing outside traditional manufacturers; the number of advertisements for senior managers was at its highest level since 1966. But the figures showed where the real jobs were coming from: while the 4,264 mechanical engineering vacancies in May were 73 per cent up on the year before, 108,000 service sector vacancies were being advertised in the same Job Centres. They also showed how the economic recovery had yet to spread beyond the South-East. Vacancies in London and the South-East grew by 20 per cent in the year to May, while in Scotland they fell by 15 per cent. By September 1985, the gap had widened further. An employment and wealth rating by Newcastle University Centre for Urban and Regional Development Studies put Winchester at the top and Consett at the bottom. Forty of the top fifty towns were south of a line from the Wash to the Bristol Channel, with the high tech towns of Bracknell, Maidenhead and Basingstoke in the top ten. The bottom was crowded by former mining, shipbuilding and steelmaking towns. Overall prosperity in the south was 30 per cent higher than in either Scotland or Wales.

As early as 1984, skill shortages were becoming a major problem. Only 20 per cent of companies in electronics were happy with the labour they could find; a CBI survey found shortages in a wide range of skills. Tom King, the employment secretary, called the problem "a quietly ticking time bomb that threatens Britain's economic recovery". The government launched an adult retraining programme at the end of the year: it gave £25 a day to small firms for every person they were prepared to train. High technology companies had the worst problems: Marconi tried to entice back electronics experts who had left the country in the 1960s and 1970s, and ran a programme to turn arts graduates into information technologists. The 1981 university cuts had not helped: the number of new computer undergraduates in 1982 was 15 per cent down on the previous year. Traditional companies were suffering as well: Jaguar was having enormous problems expanding its engineering department by 300.

Some companies were affected by that great external force, the American economy: with its huge budget deficit and Rambo dollar, it was sucking imports in like a vacuum cleaner. In 1984, Jaguar sent 18,000 cars across the Atlantic, and in the first ten months of that year, Britain's sales to the USA rose by 38 per cent.

But with British imports taking only 4 per cent of the US market, narrow segments of industry, rather than the whole economy benefited. Jaguar's success was the most visible but, as the pound tumbled towards "parity" (a dollar a pound), which it almost reached in March 1985, two other industries were having a high old time.

The civil aviation industry was in any case being pulled out of a nosedive by the deregulation of American airlines. During the recession, the number of people employed by members of the Society of British Aerospace Companies had dropped by 20 per cent, to 160,000; it would have been a lot worse had it not been for military work. At first sight, the British airliner seemed to have gone the way of the motorbike – after the triumph then the disappointment of the Comet, no more than modest success had been achieved by the VC10, the Trident or the BAC 1-11. Concorde would never pay for itself fully, but most people were at least glad it had been built. By the early 1980s, no mainstream airliners were being made in Britain.

In March 1984, the government grudgingly agreed to give £250 million to fund British Aerospace's 20 per cent share in the European Airbus, confirming the UK's permanently junior role in full-scale airliners. But that was not the end of the story: the aircraft industry had been beavering away on a new niche – small commuter aircraft – and it was these that US airlines needed after deregulation. With sterling on its back, they leapt at the bargain basement British planes. There was an upsurge in demand for British Aerospace's Jetstream commuter aircraft while the 146, which had cost £300 million to develop and was reputed to be the quietest jet in the world, came into production at just the right time. BAe was far from being the most efficient company around: despite having been privatised in 1981, it had been cushioned by defence contracts and was still an unreconstructed mishmash of the companies that had formed it in 1977. But for the moment that did not matter. It had good products, and the dollar was being kind. Britain's share of the civil aviation market stood at 17 per cent, against the 10 per cent of the 1960s.

The textile industry was a more deserving recipient of the exchange rate break. It has struggled as hard as any industry to restructure itself, and was not being comforted both by the high dollar and by a shift in buying patterns. Although the textile and clothing industry had lost 35 per cent of its workforce since 1980, it still employed 500,000 people and was, in 1984, Britain's fourth largest manufacturing industry: employment levels had stabilised and companies were even starting to invest. Apart from the falling pound, a change in the market was favouring local manufacturers. In the 1970s, jeans and T-shirts ruled: they could be churned out in vast uniform quantities by Far Eastern factories. Now, tastes were changing: old fashioned fashion was coming back (thanks in large part to Laura Ashley), and there was constant pressure for new lines. Next, the fast-growing mass fashion chain, was particularly dependent on rapid adjustment. It was impossible for companies half way round the world to react fast enough, and the exchange rate was in any case eroding their cost advantage. So British suppliers started to invest in machines that could, for example, change the pattern on a sock in fifteen minutes, instead of the ten hours it took before. New textile factories started to appear: Vantona Viyella opened a spinning plant in

Lancashire; Manchester-based Wills Fabrics spent £10 million on a spinning and weaving mill in Rochdale. Life seemed to be flooding back to a terribly bruised industry.

In 1985, for the first time in many years, exports of textiles increased faster than imports. While the value of clothing and textile imports were double those of exports, the gap was no longer widening inexorably: British textile companies had a chance to complete their restructuring without having to run backwards at the same time.

The rising dollar was not good news for all industry. British Steel was paying more for the £1.5 billion a year it spent on mainly dollar-denominated commodities. Metalbashers that relied on metals like zinc, also priced in dollars, were also hit hard.

More fundamentally, the strength of the dollar put America's free trade philosophy at risk: nothing would be more dangerous to economic recovery than a bout of protectionism, and US lobby groups were already pressing for restrictions. The danger was that European companies would be caught in the crossfire as Washington took action against the Far East. In 1984, Japan had a trade surplus of $43 billion, threatening, some said, to unbalance the system of free trade.

Not that Europeans were critical of all 300 protectionist Bills, mostly aimed at Japan, that had been tabled in the US Congress by early 1985. Most were designed to force Tokyo to liberalise its import restrictions, which were quite often bizarre. Trebor could not sell its sherbet lemons in Japan because the yellow might damage the eyes of Japanese children. Shoes were not allowed in because a politically important tribe of untouchables called the Burakumin relied on leather-making for its living. Jaguar had problems when it changed the symbol on its headlamp switches. Mrs Thatcher was herself roused to tick prime minister Nakasone off in May over the contract to build the second Bosphorus bridge in Turkey. She was convinced that the Japanese had used unfair chunks of cheap credit to snatch the job away from the British. In fact, the British were matching the Japanese terms – but the bridge provided a useful political focus.

Most of the bugbears with Japan were ingrained either into the system (a vertically integrated distribution chain discouraged shops from selling foreign goods) or into the psyche of officials. Both would take some time to change. But there was one thing that could be done: the yen, which had been kept artificially low to help Japanese exports, could be forced up. In September 1985, the Plaza Accord was signed in New York by the five major industrial countries, known as G5. Its aim was to force the dollar down, and therefore the yen up. It was to prove far more effective than any amount of anti-Japanese hectoring.

If the American economy was strong enough to do wonders for some British companies, the same could not be said about more traditional markets. The revenues of OPEC countries had fallen, not because the oil price fell (it did not collapse until 1986), but because oil-producing countries were trying to keep supply and demand in balance by rationing their own production. New production by non-OPEC members, especially Britain, was making the balancing act more tricky. Nigeria, traditionally one of the UK's biggest customers, was selling half as much oil in 1984 as it was at 1980. In addition, a whole string of developing countries, mainly in Latin America but including the Philippines and Indonesia, were finding it impossible to repay their massive loans. The world's bankers, who had made the loans in the first place, spent their time rescheduling and re-rescheduling debts. British exporters had to rely on the government's Export Credits Guarantee Department to assure them of payment, and one by one, ECGD withdrew cover on their debt-ridden markets. Nigeria owed ECGD £50 million, and no amount of hacking and trekking by intrepid exporters could sell there now. Complex countertrade arrangements, by which goods could be swapped for goods, made little impression on the overall problem.

The Tories' philosophy was one of small government – they aimed to separate state and business as far as they could, and to act positively only to break down monopolies. In February 1984, they continued their assault on union monopoly with a third piece of legislation, the Trade Union Act. Under it, unions had to hold secret ballots before calling a strike, and could be fined if they did not. In October, Austin Rover tested its powers to force the end of a pay strike. Three unions, the AUEW, UCATT and TASS, disavowed the strike, but the giant TGWU not only refused to call the strike off, it also tried to avoid paying the £200,000 fine imposed by the High Court.

Just as the hardline monetarist stance advocated by Sir Keith Joseph had crumbled as one monetary target after another was missed, so the anti-interventionism he had advocated became heavily diluted. Ministers did not start throwing money around – the chancellor's constant striving to cut public spending stopped that – but intervention for overtly political reasons became more common. Patrick Jenkin's refusal to let MacGregor close Ravenscraig was a sign of what was to come; British Leyland ran into the same problem when it wanted to close its Bathgate truck factory just before a by-election.

Not that Mrs Thatcher had made much attempt to keep her own hands off industry: state companies, in particular, were far too useful political tools to be ignored. During the 1980 steel strike, as well as during British Leyland's disputes and, of course, the miners' strike, claims that the government was letting management run the show rang hollow. Michael Edwardes complained that while at British Leyland he received constant instructions direct from 10 Downing Street.

There was another, subtler, reason why the Tories tended to stick their fingers into the pie. Ray Horrocks, who worked near the top of British Leyland for eight years and ended up running the car division, found Conservatives "much more interventionist than Labour". This was, he said, "because many Tories consider themselves businessmen just because they've run small companies". The two secretaries of state he found most in tune with industry were Cecil Parkinson and Norman Tebbit, the first men to run the newly merged Department of Trade and Industry after its creation in 1983.[1]

The knotty problem of regional policy became more acute as the North-South gap widened. A third of the DTI's overall spending went on regional aid, and the government was trying to squeeze it. At the end of 1984, Tebbit cut the total amount available and shifted the emphasis from automatic to selective help. He did at least put one anomaly right – the West Midlands, which was in such a sorry state that a junior minister had been appointed to help it in 1983, was finally given assisted status.

There had been no specific commitment to privatisation in the 1979 manifesto: indeed, the word did not then exist. Ministers thought instead in terms of giving state industries rigorous financial targets, and letting them rot if they failed to meet them. This policy crashed to the ground in the 1980 recession, with British Steel, British Leyland and ICL all being propped up with hundreds of millions of pounds of public money. Nevertheless, there were minor members of the state's lame duckery that could be offloaded with ease. They included most of the NEB's holdings: Ferranti, which had been turned round under the Board's tutelage, Fairey Engineering (which ended up belonging to Williams Holdings) and Cambridge Instruments. This had been set up in 1968 with £4.5 million of state funds to make sophisticated instruments in one of the state's rare attempts to play entrepreneur. In 1979, after it had absorbed £9 million with little to show for it, the NEB brought in a variation of the Scots-American MacGregor, a Welsh-American called Terence Gooding. With a little of his own money, and a lot of the NEB's, he turned the company round with almost embarrassing ease.

The first manufacturing flotation was part of the British Aerospace's equity, in 1981. This was an easy one – BAe looked, at that stage, incapable of making losses. But in September 1982, the government indicated that it was setting off down a headlong path to privatisation. "No industry," said Nigel Lawson, the energy secretary, "should remain under state ownership unless there is a positive and overwhelming case for it doing so."

Jaguar's flotation in July 1984 was the first fruit of this policy in the manufacturing sector. It was by far the most profitable part of BL (it made £100 million that year), and the sale would mean that the rest had to run that much faster. The government raised £300 million, but would not let the free market winds cut too sharply into its newly released pet: a "golden share" allowed it to veto any takeover bids until the end of 1990. As profitability was a prerequisite for a successful flotation, the government was always on the look out for private sector buyers for its more troublesome charges. In the same summer that Jaguar was privatised, the Department of Industry managed to sell Inmos to Thorn-EMI for £95 million, while ICL, which had received so much state money, was absorbed by the telecommunications giant STC.

There were four big ducks left: British Steel, British Leyland, British Shipbuilders and Rolls-Royce. Of these, two looked as though they could be soon on the discharge lists. British Steel managed to lose an impressive £409 million in the year to March 1985 – but that was a whole lot better than the previous year's £900 million and £180 million of it was attributed to the miners' strike. The steel market was turning up, productivity was approaching the best world levels, and the company that Norman Tebbit had called a "squawling baby" in 1981 had turned into something of a prizefighter. Rolls-Royce had been hit with the rest of the aerospace industry in the early 1980s, dipping into loss in 1982 and 1983. But, thanks to its resentful but remarkably generous parent, the government, it had been able to continue with an investment programme which was turning it into one of Britain's most advanced manufacturers.

That left British Leyland and British Shipbuilders. The car company's volume division, Austin Rover, was now the undisputed squawling baby. However much the government fed it with investment, it seemed incapable of behaving itself. Even more frustrating, it would give repeated signs that it was about to quieten down, then slip surely back into its antisocial ways. In 1983, Austin Rover managed to make a £2 million profit, compared with the £103 million it had lost the year before. That year it launched the Maestro, and the next year the Montego, cars that should have been able to earn much bigger profits than the successful but cheap Metro. Again thanks to government backing, it had been able to invest in the best equipment money could buy, and it no longer suffered from restrictive working practices. It still had strikes, but nothing like as bad as before, and compared with Ford it was a model of industrial peace. Yet in 1984, Austin Rover slipped back into the red.

It had several problems. Although there had been a surge in demand in Britain, European car sales as a whole had not budged – and the big car companies now thought on a European scale. The continent's carmakers had overcapacity of 2.5 million cars a year – five times Austin Rover's production. The Japanese increased the pressure with their increasingly attractive models, the sales of which were held back only by quota restrictions. But, rather than cutting their capacity, the big six – Renault, Peugeot-Citröen, Fiat, Volkswagen, Ford, General Motors – decided to slog it out in a bloody price war. In 1984, they lost almost £1 billion between them.

Austin Rover was in a dreadfully weak position. It was no longer a big car company. Its production was less than half a million, against the 2 million that each of the big six churned out. Car making was an industry where economies of scale were great and, despite flexible manufacturing systems that countered them to an extent, Austin Rover had to fight to make profits selling cheap cars in BMW volumes. The Maestro was a flop: it fell between two sizes in the fleet market and, in an object lesson in how not to market a perfectly good product, soon found itself labelled as a granny car. The Montego, aimed directly at fleet buyers, was dogged by a reputation for unreliability, as well as all sorts of images that it did not deserve. The curse of British Leyland was cast firmly on both cars.

In Britain, the price war was triggered by the launch of the new Vauxhall Cavalier (a rebadged German Opel put together at Luton). It was an unexpected success, and pushed Vauxhall's share of the market up from 8 per cent in 1980 to 16.5 per cent in 1985. General Motors' financial muscle was used to offer all sorts of incentive schemes to dealers, which were then countered by the equally muscular Ford, which was trying desperately to establish its Sierra as a worthy successor to the Cortina. An extra element of competition came from Peugeot Talbot, formerly Chrysler, formerly Rootes, which had been turned round by a combination of skill, money and good fortune by George Turnbull, formerly of British Leyland. Its new strength came from the phenomenal success of the imported Peugeot 205 although, by 1984, it was also assembling the new 309 in a modern factory in the Midlands.

It was not surprising, then, that in the last month of 1984 British Leyland's share of the UK market went below 12 per cent for the first time. Nor was it surprising that Downing Street was stepping up its demands that the squawling baby be dissected and sold off to whoever was foolish enough to take the bits. What was rather startling – and said much for the persuasiveness of the management as well as the softer line of the government – was that in the summer of 1985, ministers approved BL's £1.8 billion investment plan, which would include the development of a completely new engine.

For British Shipbuilders, the lame duckery was a hospice – the only question was how much of it could be salvaged. The man the government brought in to oversee this process was Graham Day. He was yet another outsider, a neatly bearded, somewhat dry, 52-year-old lawyer from Nova Scotia; but comparisons with the other outsiders, MacGregor and Edwardes, were misleading. He had never been a quick-fix man, indeed his British experience had been solely as a manager in the shipyards.

Day had first been spotted in the late 1960s when he had been sent over by Canadian Pacific to sort out a troubled contract being undertaken by Cammell Laird in Birkenhead. British officials were so impressed by his negotiating skills that in 1971 he was invited to run the struggling yard. He secured its future and in 1975 was made chief executive designate for the planned British Shipbuilders. But he became so frustrated with delays over nationalisation that he went back to Canada as a professor of business, before being lured back by the Tories in 1983. By then, of course, BS was no longer the great state-owned engine that had been envisaged in 1975. But Day knew all about rescuing British shipyards, and set to with a will.

The profitable warship yards were not a problem: the government had decreed that they must all be in private hands by March 1986. Merchant shipyards were another matter. Promising no yard was safe, Day sold off Scott Lithgow to Trafalgar House, and forced a productivity programme on to the unions. The shipbuilding force was cut from 14,000 to 10,000 with hardly a murmur, and losses were cut from £160 million to £25 million. The yards were computerised, the "who-does-what" disputes of the past were well behind but, with world orders the lowest since 1974, the future was hardly rosy.

12

# Low times for high tech

At the end of 1984, high technology lost its glow of contentment as it dipped into its own recession. It had stormed through the 1980 and 1982 slumps without stumbling – if manufacturers invested in anything, it was in computerised equipment. The M4 corridor, running west from London to Newbury, had fostered a myriad of small computer companies, as had the college-backed hothouse environment of Cambridge. Many of these companies were run by the "micro-millionaires", men mostly under 40 with great technical expertise who had broken away from the likes of ICL or Texas Instruments to set up on their own. Thanks to the Unlisted Securities Market, set up in 1980 specifically to allow small companies to float, fifty of them had already made their first million, on paper at least.

In Scotland, a different type of high tech company was arriving. The efforts of the Scottish Development Agency in tempting investors had taken some of the pain out of the collapse of engineering and shipbuilding: by 1984, 42,000 people worked in 270 high tech companies. The SDA claimed to have the second highest concentration of silicon chip makers in the world, IBM had installed European production of its standard Personal Computer at Greenock and in 1982, Wang had invested £40 million in a computer factory on the campus of Stirling University. A handful of homegrown companies had also flourished, notably Glenrothes-based Rodime, which introduced the first ever three and a half inch computer disk drive, later to be adopted by IBM as its standard. "Silicon Glen", running right up the country

# Low Times for High Tech

from the Clyde in the west to Aberdeen in the east, was one of the success stories of the early 1980s.

But at Christmas 1984, the home computer boom stuttered, triggering a downturn in the whole sector. The enthusiasm with which British children had taken to computers was one of the more hopeful signs for the future: in 1982, Britons had twice as many computers per head as Americans and one and a half times as many as the Japanese. The government had sponsored a programme to put a microcomputer into every school. At Christmas 1983, shops had found themselves running out of machines, so they packed the shelves for 1984, assuming demand would be well up on the previous year's record level. They were wrong: no one had noticed that the market was nearing saturation. As 300,000 unsold machines gathered dust, computer companies resorted to savage price cutting to keep their factories working.

None was hit more severely than Acorn, a Cambridge company that had won the contract to build the standard BBC computer in 1981, and subsequently approval for it to be used in schools. Turnover had shot up from £9 million to £42 million in a year and at its peak, the Unlisted Securities Market had valued Acorn at £136 million. But the founders, Herman Hauser and Chris Curry – two of the best-known micro-millionaires – though technically brilliant and tough negotiators, lacked financial discipline. They diversified into biotechnology and satellite communications, and even sponsored a Formula 3 racing team. Their launch in the USA, in 1983, lost £6 million and, when the Christmas 1984 débâcle came, their computer lost out to Clive Sinclair's cheaper models. With more than £4 million drained by the Christmas advertising campaign, they found themselves heading rapidly towards receivership. In February 1985, Olivetti moved in and took a 49 per cent share in Acorn for £10 million; still the management could not sort itself out, so the Italians decided to take complete control, bumping their share up to 80 per cent.

As home computer sales slumped, the microchip market went into a nosedive. The silicon chip had already established itself as the ball bearing of the 1980s – every new product seemed to have at least one and, not surprisingly, companies had poured investment into factories to make them.

The chip market was notoriously susceptible to violent business cycles. After a 40 per cent rise in demand in 1984, it collapsed by at least the same amount in early 1985. Japanese makers started a chip war, slashing prices, and manufacturers everywhere saw their profits evaporate. The slump was not just a British phenomenon, but when Silicon Valley caught a cold, Silicon Glen got pneumonia: too many factories were vulnerable subsidiaries of American groups. In June 1985, chipmaker National Semiconductor announced it was laying off 450 workers at Greenock – a year after it had said it was going to invest £100 million to create 1,000 jobs.

In the summer of 1984, Thorn-EMI had bought the government's majority share in the chip manufacturer Inmos for £95 million. By the year's end, it was looking like a good deal as Inmos announced a £14 million profit, compared to the £13.5 million it had lost in 1983. Then came the chip crash, and by spring it was again trading at a loss. The company's original aim, to produce standard "commodity" chips, was abandoned, in favour of specialist products; but its great hope, the transputer, which combined the main functions of a computer on one chip, had yet to be launched. In the summer, jobs were cut at both the Newport and Colorado plants.

The downswing continued throughout the whole high tech field. To an extent, this was a function of the growing power of the City, which created a mood that invariably exaggerated the real state of business. This was "sentiment", and it was picked up and passed on by journalists.

Nevertheless, there were signs that British high tech companies were starting to follow a familiar path. In 1978, electrical and electronic engineering ran a trade surplus of £700 million; by 1984, it was £1.4 billion in deficit – the worst performance of any sector except vehicles. In the same period, the office machinery and data processing sector (which included computers) increased its deficit from £200 million to just over £1 billion.

Information technology, or the linking of telecommunications and computers, was one area where the government had intervened with something like zeal. Even Sir Keith Joseph had approved a scheme by which the Department of Industry would help companies raise the money they needed. Kenneth Baker had been made Minister of Information Technology and 1982 had been declared Information Technology Year. IT grew rapidly, finding its way into practically every other industry, and by 1985, the world market was reckoned to be worth $300 billion and to be growing at 15 per cent a year. But, said a report by an NEDC committee in 1984, "We're talking about a sunrise industry that is being eclipsed before it had properly risen."

Part of the problem was that the government preferred not to invest directly in IT, but to encourage its expansion by, for example, lifting restrictions on cable television operations. But industry responded creakily to these opportunities: the NEDC report cited familiar problems of poor marketing, lack of investment and a shortage of skills. It was not the whizzkids that Britain lacked, it was "journeymen technologists", who could apply the new technology.

A sideshow to the high tech extravaganza in the first half of the 1980s was the story of biotechnology. It had been the exciting sister of IT, at least as beautiful and certainly more exotic. Biotechnology was the manipulation of living cells in manufacturing. Fermentation was the simplest form but, scientists were saying in 1980, it could be used for activities that ranged from metal extraction to a male birth control pill; genetic engineering was its most controversial use. Investors, desperately looking for new places to put their money, had leapt into the biotech pond. But after two years of shaking test-tubes, products stubbornly refused to come out of them. Even though one British company, Celltech, was moving tentatively forward, it became increasingly apparent that, while biotechnology would become important, it would not be in the 1980s, or possibly even the 1990s. The little sister bowed out – she would take a long time to mature.

The Tories' free marketism was providing a bracing new climate for the electronic giants like Plessey, STC, GEC, Thorn-EMI and Racal, all of which handed in poor results for 1984. Their problems came principally from the loss of steady contracts, especially from the Ministry of Defence and British Telecom. The ministry had become much tougher with its suppliers, while the newly privatised Telecom was showing a worrying willingness to buy abroad.

This nonplussed their traditional contractors, who had continued to rely on links with Whitehall and Westminster as the free market did its worst with less fortunate sectors. In 1984, four customers – the Ministry of Defence, British Telecom, British Gas and British Rail – accounted for 25 per cent of GEC's turnover. Lord Weinstock had always packed his upper ranks with retired generals, admirals and top civil servants. The nature of the "cost-plus" contracts meant that GEC could not fail to make money; and of its vast research and development budget – £620 million in 1984 – only a third was borne by the group itself: customers paid for the rest.

The Thatcherites were offended. Peter Levene, a hardnosed former chairman of United Scientific Holdings, was put in charge of the Ministry of Defence's procurement section and soon started to irritate Weinstock. But GEC did itself few favours by failing to deliver the Airborne Early Warning version of the Nimrod reconnaissance plane. This had been ordered by the Labour government in 1977, even though the RAF had expressed a preference for Boeing's AWACS, and in July 1985, with £900 million already spent on development and the plane running five years behind schedule, Levene ordered a review of the whole programme. Weinstock's irritation was being reciprocated.

The heavy end of high technology was, however, rather different from most industries. For one thing, it was by its nature largely dependent on public sector customers; for another, it was the most international of all businesses; and for a third, it needed such constant massive infusions of development money that the economies of scale were virtually limitless: the higher the sales, the more R&D costs could be spread. Government attempts to ensure free competition within Britain were putting the electronics majors at a disadvantage to their competitors – even the supposedly free-marketist American government vigorously supported its own industry with defence contracts.

Guaranteed government and state industry contracts had given the big companies a steady home market from which they could expand. Even so, no British company had reached anything like the size of Siemens of Germany, or AT&T of America, which were able to grow on the back of much larger and more dynamic economies. Research costs were as high, but they had to be spread over smaller guaranteed sales. The Post Office had tried to overcome the problem in the late 1960s when a new telephone switching system, System X, was conceived. It would be built jointly by GEC, Plessey and STC so that the costs could be shared. Bickering and horse trading between the three slowed development (STC eventually dropped out), and the hoped for export sales failed to materialise as more fleetfooted rivals whisked contracts away. By the mid-1980s, the installation programme in Britain was finally moving, but exports consisted of one £2 million sale to the island of St Vincent. Costs per line remained way above American levels, making it impossible to sell more systems: it was a vicious circle.

As they struggled to find a role in the new freer market, the electronics giants had started to shake themselves up – albeit not as violently as their smokestack cousins. In 1980, Ernest Harrison's Racal had bought Decca, a once dynamic company that had grown arthritic as its founder, Sir Edward Lewis, had failed to let go of the reins. Harrison was a new boy: he had joined the infant Racal as its accountant in 1951, and had built it up into a £500 million electronics group. In the face of a counter-bid from GEC, and with the eventual blessing of the near octogenarian Sir Edward, Harrison had won the prize. His aim, he said, was to make the British electronics industry internationally competitive.

Since then, Harrison had been moving Racal away from defence hardware, concentrating particularly on mobile telephones. But that had not persuaded the City that it should be spared the hammering the electronic sector was receiving: Racal was one of the seventeen high tech companies that were among the thirty worst performers in the share index in early 1985. Its problems were as nothing, though, compared with those of two other companies – STC and Thorn-EMI.

In the early 1980s, the telecommunications giant STC had been thoroughly churned up. It had shed its majority shareholder, ITT, it had dropped out of System X, and in 1984 it had taken over ICL. Thanks to Wilmot's upheavals, the computer company was now performing reasonably well but the rest of STC was being hit hard by the change in government procurement policies. In 1985 it scrambled to adjust by cutting 2,300 jobs and axing some of its more mundane businesses.

Even as the group was moving belatedly in the right direction, chairman Sir Kenneth Corfield made the worst mistake he could. An electrical whizz (he tried to build a television set at the age of 10), he failed to understand the importance of impressing the City. In March, he talked of "an encouraging outlook", and the institutions supported a £168 million rights issue. But the share price started to tumble and in July, when Corfield announced that STC had made a loss in the first half of the year, they started baying for his head. In September they got it, when he and five other directors resigned. The new chairman was Lord Keith of Castleacre, an accountant, architect of the merchant bank Hill Samuel and known as one of the hard men of British business.

In May 1984, shortly after he had taken over as chief executive, Thorn-EMI's Peter Laister announced that he wanted to merge with British Aerospace. Laister had arranged the acquisition of the music and medical electronics company EMI (which had been dragged down by the Nobel Prize winning but financially draining invention, the body scanner) in 1980, and saw nothing odd about combining a music and television company with a major aircraft and defence contractor. The deal never came off, so Laister turned to Inmos. The City went on its guard: "sentiment" was turning against mergers that had no apparent industrial logic.

In 1985, the ambitious Laister was in real trouble. Not only had Inmos turned sour, the Ferguson television group was struggling after failing to adjust from large to small screens and the American record division was running into trouble. In June, Ferguson announced 1,000 redundancies and Laister was thrown out in a boardroom coup: his successor criticised him for thinking too much about strategy and not enough about day-to-day problems.

Now the future of Inmos, one of the few British companies with a world technological lead, was in jeopardy: would the City let Thorn feed it the money it needed? The *Guardian* summarised the problem when the transputer was finally launched in October 1985: "Thorn-EMI might not risk an attack of City vapours (and the risk of an unwelcome bid) by investing the scale of funds necessary for a full-blooded attack on world markets, given the added risks of being first with the transputer." That summarised a problem that was now being discussed more widely: City short-termism. How could a creature like Inmos ever survive as part of a British public company when pension and investment fund managers were having their performance monitored by the quarter or even by the month? The new management at Thorn knew the answer – it decided to offload its precocious child as soon as it could.

Of all the high tech disaster stories in 1985, none was as public as Sir Clive Sinclair's. By the summer, Sinclair Research had debts of £15 million and £34 million worth of unsold computers in stock. And the little open-topped electric tricycle that he had launched in freezing January, the C5, was popular with cartoonists, but nobody else.

If Britain had too few journeyman technologists, it also lacked entrepreneur technologists. The business wizards – the likes of Owen Green and Lord Hanson – kept as far as they could from the complexities and uncertainty of high technology. Unfortunately, high tech wizards, although filled with enthusiasm and energy, rarely found it possible to buckle down to the mundane business of running companies. Too often they were chip brains – speeded-up versions of the old spanner brains.

It was a shame, because Britain had no shortage of high tech wizards. The British education system was terrible at producing people who created wealth, but great at generating geniuses. The lack of structured university teaching allowed people with the ability to follow their own path unhindered: Cambridge's Trinity College had, after all, produced more Nobel laureates than France.

In only one area, pharmaceuticals, had the British shown that they were capable of moving relatively smoothly from the laboratory to the production line. Glaxo's Zantac was now the best-selling drug in the world, and other pharmaceutical companies – led by ICI – were pumping vast quantities of money into research.

But the drugs companies had their scientists, who could then be tamed by conventional managers. Much of the newest electronic technology and computer software was developed and marketed by scientists who had decided, wrongly, that they could also manage. Their epitome and king was Clive Sinclair.

At the age of 10, Sinclair's prep school masters told his father that they could not teach him any more mathematics. He invented the binary system on his own (without realising someone else had done it 200 years before). And he could not bear to see anything without finding out how it worked. He was a genius with an obsession with electronics, and an unstoppable capacity for inventing things.

Having decided university would stifle him, he left school at 17 and worked as a technical journalist. His first business was packaging and selling kits for cheap hi-fis and radios, including one the size of a matchbox; in 1966, he announced the first pocket television, although it was never put into production. Sinclair was obsessed by compactness, which frequently led to unreliability – it was quite normal for one of his products to be replaced several times before it worked. In 1972 he hit the headlines with the world's first pocket calculator: it was technically brilliant, but when the Japanese copied his techniques and then flooded the market with cheaper and improved versions, he did not fight back with a new calculator but launched instead the Black Watch, the first digital wristwatch. It suffered from component problems, and Sinclair had to be bailed out by the National Enterprise Board. In 1977, the NEB put £650,000 into his new pocket television – it too flopped. The television, he said, would be bought by executives: it wasn't. Again and again, he paid the price for impatience: markets were not properly researched, bugs were not ironed out before a launch, and Sinclair's reaction to falling sales was just to launch a completely different product.

In 1979, having lost his business, house and Rolls-Royce, he split with the NEB, which wanted him to concentrate on his successful but unexciting digital meters, and set up a new company, Sinclair Research. With his first product, the ZX80 – the first computer for less than £100 – he hit gold, and with its successor, the Spectrum, he hit platinum. It sold 200,000 in its first nine months and, as the world's best-selling computer, gave his company a £14 million profit in 1983. He went back to miniature televisions, and came up with a tiny flat screen television. Unfortunately, due to problems with development, and with the Timex factory in Dundee where it was to be made, he was pipped to the post by Sony. That probably did not matter much – it turned out very few people wanted to watch tiny televisions anyway.

In 1983, when the Spectrum was selling wildly, Sinclair cashed in part of his company to City institutions for £13.6 million. He used the money to develop the C5, which was launched in the middle of a chilly January in 1985. It was supposed to replace petrol engined cars in cities, but was so small and vulnerable that road safety organisations condemned it, and hardly anybody bought it. Production at the Hoover factory in Merthyr Tydfil where it was made was cut back from 1,000 a week to 100 in April: Sinclair had pumped £7 million of his own money into it. He admitted that he thought it was not a very good design, but hoped it would fund the development of a full size electric car. Unfortunately, it did not even fund itself.

Although the Spectrum had survived the disastrous Christmas of 1984 better than most, retailers put in next to no orders in the next two months as they tried to clear out their stocks. As a result, Sinclair's own stocks started to pile up in the warehouses. By the summer, it became clear that he would have to find money from somewhere. A number of big companies looked and backed away, alarmed. Then Robert Maxwell, the larger than life printer and publisher, announced on the front page of his *Daily Mirror* that he would come to the rescue of this other great Englishman. His accountants took a close look at Sinclair's books: in August, a small story on page two of the *Mirror* announced that Maxwell had changed his mind.

Sinclair had a whirling top of a brain, but one that was remarkably difficult to tame. His experience with the sober-suited NEB men taught him that he did not like sharing power, but he then appeared incapable of running a business successfully. He well understood the importance of marketing – he was his own best sales gimmick – and his balding, bespectacled figure became one of the best-known images in the country. But he found figures tedious, and always tried to substitute enthusiasm for a grasp of detail. He was, in fact, the complete antithesis of the bean-counters who were now methodically tidying up industry. His skills were desperately needed, but there seemed no way of fitting them into the British business machine.

Alan Sugar, by contrast, fitted the machine perfectly. He belonged to that rare breed for which Sir Keith Joseph had cried out, the natural entrepreneur: he was a tough, instinctive, wheeler-dealer who always worked from the market backwards. He had started selling car aerials in his native East End of London, had graduated to cheap but slickly packaged hi-fis, and had floated his company on the stock exchange in April 1980. In 1984, he launched a cleverly packaged games computer, the CPC464, which cut through the downturn in sales to take 10 per cent of the personal computer market within six months.

The next August, he launched his all-conquering PCW8256, a word processor good enough for small businesses that cost less than £400. The City adored Sugar, but he did little for the British manufacturing base. One of his skills was finding the cheapest place to have his computers manufactured: invariably, that meant somewhere in the Far East.

In April 1986, when Sinclair's computers were still taking 40 per cent of the British market, Sugar bought the Sinclair name for £5 million. "We're good at initial marketing," Sir Clive commented at the time. "But we're not in the same league as Alan Sugar when it comes to mass marketing." He went back to what he was good at – inventing things.

Mark I'Anson, co-founder of Integrated Micro Products, did not have the firecracker brain of Sinclair nor the business genius of Sugar, but he did have a combination that was at least as rare: a high-powered scientific mind combined with the patience to learn about the mundanities and frustrations of business. On his journey from science to entrepreneurdom, frustrations were a major feature.

As soon as IMP had started production in 1982, it had become clear that its sophisticated computer boards were going to be a winner. Universities and large companies queued up at the windswept door of the Consett factory and within a year sales had generated the £100,000 needed to repay the company's overdraft. By this time, life for I'Anson had become a constant search for funds. Still helped by BSC Industry (which threw in its own £20,000 unsecured loan), he persuaded the British Technology Group to put in £50,000 as equity. BTG was the rump of the National Enterprise Board, now a high tech venture capital organisation. Whenever the IMP bought capital equipment, it also received a 20 per cent regional development grant; as the major capital expenditure was on desks, that sum was limited. Still it had virtually no private sector backing.

In 1984, I'Anson discovered what rather too many people knew quite well: that the real way to make money was to do a deal. The government had decided that the BTG should concentrate on technology transfer, and should shed all its venture capital activities. The timing could not have been better for IMP: it was running at maximum output and the next tranche of development expenditure had not yet started. It bought BTG's shares and sold them on to two venture capital groups for a £200,000 profit.

But then came the high tech slump, just as the company was gearing up to produce its next generation of computer board. IMP slipped and slithered into loss, not least because it was spending 50 per cent of its turnover on research and development. I'Anson went to his bank and asked for an extra loan. The bank offered half what he wanted and, to his astonishment, suggested that he would have enough if he trimmed his R&D spending. In the end, he took what he was offered and managed to scrape through until the market turned up. Somebody, he considered, should tell the banks that there was supposed to be an enterprise culture.

In December 1985, GEC bid for Plessey. It was Lord Weinstock's answer to two problems: first, how to create a company big enough to compete internationally; second, how to spend the £1.4 billion of cash that his careful husbanding had built up over the previous two decades.

GEC was the great steamroller of the electronics industry. Before the War, it had grown up as the empire of one man, Hugo Hirst, and by 1939 employed 40,000 in light electrical engineering. But after Hirst's death in 1943, the group lost its way, and it was not until 1961 that it found it again. That was when GEC took over Radio and Allied Industries, a company controlled by Michael Sobell and run by his nephew, Arnold Weinstock. Soon, Weinstock was in charge of GEC, and was bashing it into shape. He cut the headquarters staff from 2,000 to 100, and imposed his own austere variety of financial control. In 1967, he embarked on a great restructuring of the electrical industry. Backed by the IRC, he merged GEC with AEI and English Electric to create Britain's largest electrical group: its 160 companies made products that ranged from nuclear power stations to washing machines.

GEC became a cash machine, run in many ways like Hanson or BTR. The subsidiaries, which Weinstock rarely visited, were monitored by monthly figures. He encouraged managers to go for high margin, low volume products, and personally approved any bids or salaries above a relatively modest figure. By 1985, the bid figure was £1 million, the salary £25,000. Although many of the companies were operating at the upper end of technology, research and development spending was mostly absorbed by customers. He kept away from microchips and borrowed robot technology from Hitachi. Weinstock was, as the City said, "risk averse": it showed through in his profits for the first half of 1985 – as other giants swung into loss, GEC's earnings dropped by a tolerable 13 per cent.

But if it was not pushing the boundaries of technology forward as far as it might, GEC was one of the great contributors to the British economy. Consistently one of the largest exporters, it showed great skill in winning overseas contracts. When the British power station market dropped off, its power engineering division hoovered up projects around the world, notably in India, Hong Kong and South Africa.

What the City really hated about Weinstock was his refusal to buy other companies. By 1985, his only major purchases were AB Dick in America, and Avery in Britain; attempts at taking over Decca in 1980 and British Aerospace in 1984 were less than wholehearted.

Plessey was a company much more in the traditional mode of British business, one of the last great family-run groups. But it had at least been thoroughly shaken up by its own recession. Founded after the First World War by the American-born Sir Allen Clark, it was now run by his sons, Sir John and Michael. Sir John, educated at Harrow and Cambridge and a self-confessed paternalist, had fought off an attack from the non-executive directors and had taken charge of the company in 1962. Plessey had grown rich supplying telephone exchanges to the Post Office; when, in 1976, the Post Office announced it was cutting right back on orders, Plessey was left floundering. With System X still on the drawing board, the Clark brothers were forced into action: by 1980 they had cut the workforce from 85,000 to 47,000, and had plugged themselves more firmly into defence electronics. In October 1980, Plessey won a £150 million contract from the army for its Ptarmigan telecommunications system. But in the next few years, it did little to reduce dependence on government orders. With System X stumbling, 80 per cent of defence orders still coming from Britain in 1985, and the City unconvinced by the Clarks' moves to loosen family control, it was clearly vulnerable.

In bidding for Plessey, Weinstock would have spent £1.2 billion of his cash, and created a group that would still have telecom sales a quarter those of the world leader, AT & T. But the Office of Fair Trading recommended that the bid be referred to the Monopolies and Mergers Board, and the government agreed. Weinstock would have to twiddle his thumbs for six months while the MMC made its investigation. British merger policy was based on purely national interest; international competition was not considered and GEC knew well that it would have to fight hard to get clearance. The system took a dangerously myopic view of increasingly international businesses.

# 13

# *Laying the charges*

By 1985, the government was starting to crow a little as figures came through showing manufacturing productivity had risen by 23 per cent between 1980 and 1984: production was back to the level of four years before, while the workforce had fallen by 22 per cent. That, said ministers, proved that industry had responded to the shock treatment, and was well on the way to being fit, if slim. No, no, said its opponents, all the figures showed was how bad industry had been before. Far from being fit, it was emaciated. In 1985, average profitability – measured as the rate of return on capital employed – was 8 per cent, compared with the low point of 3 per cent in 1981. That, the opponents said, could also be explained by the general recovery.

In October 1985, the month of the Broadwater Farm riot in North London, the House of Lords Select Committee on Overseas Trade, chaired by Lord Aldington, chairman of the insurance company Sun Alliance, came down firmly on the side of the pessimists. It had been asked to find out why the balance of trade in manufactures had gone into the red for the first time in 1983, and particularly why the non-oil balance of payments had plunged from a small surplus in 1981 to an £11 billion deficit in 1984. It pointed out that the deficit had come about entirely through a 40 per cent increase in the volume of imports since 1980: exports had remained remarkably steady given the yo-yoing exchange rate. The scale of the problem could best be seen through an international comparison: West Germany's *surplus* on manufactured goods in 1984 – £50 billion – was greater than the *total value* of British manufactured exports. Alarmingly, while the worst performance had come from motor vehicles (going from a £334 million surplus in 1978 to a £2.5 billion deficit in 1984), the runner-up for the wooden spoon was high tech. Electrical and electronic engineering went from a £697 million surplus to a £2 billion deficit.

Its witnesses were, on the one side the Treasury and the Chancellor of the Exchequer, Nigel Lawson; on the other, a host of businessmen, mostly from large companies and trade associations.

The report[1] was produced against a background of unemployment that seemed destined to stay permanently above 3 million. The Lords said that unless the government did something about its policy on manufacturing industry, economic disaster would follow. They were particularly perturbed by the government sanguinity in declaring, first, that as North Sea oil revenues fell, so manufacturing would rise to compensate (because sterling would fall) and second, that it did not in any case matter whether wealth was created through manufacturing or services – it was total output that counted.

The sight of Treasury officials waving their economic textbooks irritated the committee members, most of whom were practising businessmen. They demolished both bits of theory with common sense. Manufacturing, they said, could not be turned back on like a tap: so much equipment had been sold or scrapped that it would be difficult for industry to cope with an upsurge in demand. Many companies were already operating at near full capacity and plant now being installed was, for the most part, designed to increase efficiency and save labour: only 20 per cent was to raise capacity. "There is also," they added, "the question of human capital – the skills in design, operation and marketing and the goodwill and sales contacts – which have been lost, perhaps irrevocably." They could also have mentioned, though they did not, that the Treasury's belief that there was a close inverse relationship between the strength of sterling and the level of exports had not been proved by recent experience.

The belief that services could take the place of manufacturing was nonsense, the Lords declared. First, services were much more difficult to sell abroad (only 20 per cent were tradeable): they would never be able to generate the foreign currency needed to pay for imports. Second, manufacturing was the customer base of many services – close the factories, and the services would go too. Lord Weinstock of GEC was quoted approvingly. "What will the service industry be servicing when there is no hardware, when no wealth is actually being produced? We will be servicing, presumably, the production of wealth by others. We will supply the Changing of the Guard, we will supply Beefeaters around the Tower of London. We will become a curiosity. I do not think that is what Britain is about; I think that is rubbish."

The Lords travelled to France, Germany and Japan, and identified the root cause of the problem as a cultural one. They pointed to the esteem in which industry was held in these countries, and the way that everybody seemed to pull together to strengthen it further. German training was praised (a study had recently concluded that it was responsible for labour productivity there being 63 per cent higher than in Britain). The Japanese school system, which provided a far higher level of basic education, was noted, as was the fact that there were four and a half times the number of engineering graduates in Japan as there were in Britain. The committee was impressed by the industrial strategies of other countries, which were continuous, long-term and carefully selective in their support. In contrast, the committee came across some startling shortcomings in the way British companies still did business: less than 20 per cent offered delivered (that is, all-in) prices to customers; far fewer British than foreign companies used market research; and most firms were still afraid of quoting in foreign currencies, putting themselves at an unnecessary disadvantage given the abundance of techniques available to protect against losses through currency movement.

The committee was not completely bowled over by its foreign trips. It concluded that much Japanese success was due to companies' abilities to compete fiercely within a home market protected from imports. And that even a supposedly free market country like Germany resorted to massive industrial subsidies: a special bank provided £1 billion in low-interest loans to small businesses in 1983/4 alone.

The long-term answer, the Lords suggested, was through education: an awareness of the importance of economics must be taught at school. But in the interim, solutions should be borrowed from other countries. As the countries they visited were all corporatist by philosophy, their ideas all smacked of corporatism. Macro-economic policy should take the needs of industry into account, to stop exchange rates bouncing up and down. Industrial priorities should be decided by a body involving groups like the CBI and TUC. And the government should increase subsidies for R&D, and maintain support for services for exporters.

The Lords shot themselves in the foot with their relentlessly corporatist approach: they were trying to sell whisky to a teetotaller. Their failure to look at the most powerful economy in the world, the thoroughly free marketist United States, combined with a lack of discussion of improvements in British productivity, laid them open to ridicule, which they duly received from government ministers.

But they had touched on most of the really important issues. The fundamental need was well understood by the government. Norman Tebbit, the Trade and Industry Secretary, said in his evidence that the single most important factor in achieving extra growth was a change in culture. But the disdain with which the government appeared to treat manufacturing industry was bound to cause concern. (In November 1985, the CBI published a survey claiming that the main reason for industry's problems was that the government had "written off manufacturing".) They were right to say that services could not replace manufacturing: Harvey-Jones came up with another statistic when he gave a downbeat Dimbleby lecture the following spring: it would take an extra 6 million tourists, a 40 per cent rise, to replace ICI's contribution to the balance of payments. And the Lords identified a problem that would, five years later, still not be widely understood: that British manufacturing industry was too small, and would have great difficulty in becoming big again.

Anyone reading the report, with its appendix of distinguished business witnesses, would assume that the entire business community was lined up against the government. That was not at all the case.

Most businessmen, however much they grumbled at the government and complained about its neglect of manufacturing, vastly preferred it to its predecessor. Managers no longer felt like second-class citizens: whereas the Labour Chancellor Denis Healey had said that he would squeeze the rich until the pips squeaked, Mrs Thatcher had praised Coloroll as an example for all to follow. She had pressed down on civil servants, university lecturers and other pampered paraphernalia of the past, and businessmen had automatically risen up the pecking order. She may have been following the public mood rather than leading it, but at least she was in tune with it. Many businessmen felt the same as Harry Solomon of Hillsdown: "We couldn't have done what we did ten years earlier. I feel at that time the mood of the country was not ready for companies like ours. There was a feeling of resentment against entrepreneurs and businessmen."

The trouble was, Mrs Thatcher's personality meant that top industrialists were forced into the pro- or anti-camp. Hugo Young called his biography of her *One of Us* – it was only "one of us" who would get the important political posts, but the concept held good in the business world. Even as she rejected the old establishment of suave civil servants and lawyers, she created a new establishment. Her heroes were the likes of Lord Hanson, Lord Taylor (of Taylor Woodrow), Lord Forte (of Trusthouse Forte) and Lord King (who was busy turning British Airways round). It was no coincidence that they were all lords. For the most part, they were self-made men, often from modest backgrounds like her own, and were involved either in services or the predatory side of manufacturing. Sir Owen Green would never become part of the "Thatcher mafia" (why wasn't he a lord?), but gave her tacit support. New-style managers up and down the country did the same.

Witnesses to the committee included those who were definitely not "one of us". Sir Terence Beckett, Lord Weinstock and Sir John Clark, of Plessey, made no secret of their belief that the government should be helping industry more directly; Sir John Harvey-Jones, chairman of ICI, was even worse – a paid-up member of the Social Democratic Party.

And yet it was difficult to deny that Harvey-Jones was among the most radical businessmen in the country; or that Lord Weinstock ran as tight a business as Lord Hanson. The truth was that these groupings were not really important in the real world of industry. What was important was what was going on inside companies themselves. There, the Lords failed to delve. Had they done so, their conclusions might have been somewhat different.

# Part Five

# *Shaking the iron universe*

14

# *People upheaval*

British industry had changed because it had no choice. Had it not, the worst fears of the Lords' committee would already have been on the point of fulfilment. The pressure had been building ever since the near captive markets of the Empire had been set free, and foreign companies had realised that the British market itself was ripe for exploitation. First the American multinationals had arrived, bringing their new-fangled management techniques with them; then the Germans and other Europeans had started to chip away at the British market; finally, the Japanese had come. Just as the pressures were becoming unbearable, North Sea oil, world recession and government policy had come together to trigger the cataclysm of 1980 and 1981. When that eased, the predators arrived. It was one damned thing after another and, as the Lords pointed out, something had to change. In parts of industry, it already had.

The changes started with people. Manufacturing industry had been depopulated for good – it would never again be a major sink for the British workforce. But, as the statistics were showing, it was still producing as much as it had before, partly because of automation but much more because of another human upheaval, this time among the managers. Changes at the top had led to an exhilarating ripple throughout companies which, combined with the application of new techniques for making things, did more for British industry than anything else this century.

The changes took place in two stages. First, there was the actual replacement of people. The average age of top management fell, as senior managers were pensioned off and replaced by younger executives. There were exceptions: two of the most effective 1980s-style managers, Ian MacGregor and Lord Keith, were well into their sixties when the decade started. But the generation that benefited most was the one that was born around the end of the War, and was heading towards 40. These were the people given their chance by Harold Wilson's new colleges. Many of them were bright kids from poor families, who had been catapulted via the grammar school system into universities and polytechnics. Most had some sort of professional qualification; a growing handful (still distrusted in many companies) had been to business school. Whether accountants, engineers or MBAs, they were Mrs Thatcher's natural constituents, and despised as much as she did the panoply of corporatism and anti-industrialism.

As they worked their way up through the ranks of companies in the 1970s, they could see all too clearly what was wrong with the management: the lousy financial controls, the rigid structures, the complacency. Many had worked in foreign-owned companies, and saw how these tended to outperform their British rivals, and why. When they were given a chance to do something about that, they leapt at the opportunity.

The second stage was the change in culture. Even in the worst hit companies, the vast majority of managers remained in place. But as new, more rational policies were put in place by their bosses, they responded and started to work more efficiently themselves. When Ian MacGregor took over at British Steel, the government expected him to clear the management out and start afresh. He took that as a challenge, and decided to work with the existing managers, simply giving them targets to meet. The results of a little bit of motivation soon showed up in BSC's results.

As managers realised that there was no fundamental reason why their companies should not recover, they rolled up their sleeves and started to enjoy themselves. Tony Gill, who was then managing director of Lucas, remembers the early 1980s as "an exciting time, and progressively less frustrating as we were able to do things we obviously needed to do".

How widespread were these changes? That was difficult to say. When Sir Peter Parker, ex-British Rail boss and chairman of the British Institute of Management, wrote an article in 1985 suggesting that they were extensive, the *Sunday Times* was deluged with letters from people insisting that he was living in cloud cuckoo land. There were certainly plenty of badly managed companies still around – but there were many more good ones than there had been before the recession.

The men (there were few women, if any) who were having the biggest effect on British manufacturing were mostly competent technocrats, few of whom would stand out in a crowd. They were people like Gareth Davies at Glynwed, a quietly spoken accountant who none the less pulled his company into the first rank of British industry. Or Tony Gill, an Essex-born engineer from Imperial College, one of the driving forces behind Lucas's fight back; in private amusing and sharp-witted, in public undemonstrative, even slightly shy. Or Bernard Robinson of Tallent, who left school at 15 and took himself off on an accountancy course when he reached the board. One successful chief executive took his professionalism to an extreme: he liked to relax, he said, by reading other companies' annual reports.

The new bosses had to be more than just technocrats, of course: they needed considerable force of character first to get to the top of their companies, and then to push them off on an unfamiliar course. John Harvey-Jones, with his flair for publicity, was exactly what ICI needed when he took over as chairman in 1982 – his gaudy ties were a semaphore to everyone on board that the mighty group was trying to change direction. But "characters" were the exception. Even after the media had decided that business should be a source of glamour, after every television channel had started its own business programme and a high proportion of the population had become shareholders, business leaders remained an anonymous lot. Who would the man on the Clapham omnibus recognise: Richard Branson and Robert Maxwell, probably; Harvey-Jones and Alan Sugar, possibly; ditto Tiny Rowland and Rupert Murdoch. Who else?

Yet they all had to have charisma, however quietly expressed. It was needed both to spur on their managers, and increasingly to impress the men in the City. John Waddington kept its independence because the man from the Norwich Union liked its managing director; Dunlop was sold down the river at least in part because bankers did not like Alan Lord.

The predators needed a double dose of charisma, for the same reasons. Harry Solomon would never be recognised on the street, because he hated having his photograph published – he believed it would be unfair on his children – and was furious whenever one of the few shots taken appeared in the press. Yet he was one of the most charismatic of all the business leaders, with a trick of flattering his interlocutors – "You'd make a good managing director" – and encouraging his managers to drop in for a cup of coffee whenever they were in London. Owen Green, tough as old boots underneath, was the sort of chap you felt you could spend a happy few hours in the pub with. He would ring his managers up "for a bit of industrial gossip – I think it's valuable to have a chit-chat".

It was more difficult for young men to have this charisma, but all the successful ones did their best to motivate their managers. John Ashcroft used jokiness – he put his photograph next to ten pithy slogans on mugs that he gave to managers. Nigel Rudd, altogether a shyer man, was more conventional but also more open: he was good in individual meetings, and fund managers loved him for his deal-making flair.

Running back through this short list of business leaders, it is interesting to see that all except Harvey-Jones went to grammar school, and his background (childhood in India, naval college) kept him away from the Establishment ladder. "I think the success of any capitalist country is the ability of people from low or poor backgrounds to come through," says Nigel Rudd. "The grammar school was the great opportunity for people like me to succeed."

It would take a very long time for the inheritance of a century of indifferent management to be scrubbed clean, but the new bosses started by discarding some of its more obvious symbols. They threw out executive loos and dining rooms, and they loved the press to dwell on how squalid their headquarters were. The predators were the masters of this – item one on any takeover plan was to chop back the HQ. BTR's shoebox in Pimlico was a notable example of grottiness, although Williams' base, all mixed up with a car showroom in Derby, was a good attempt. However small Hanson and Hillsdown's headquarters were, their addresses – in Knightsbridge and Hampstead – put them well down the squalor list.

But industrialists played the game too. John Harvey-Jones wanted to move ICI's headquarters out of its fortress by the Thames; in the end, it shrank itself into a part of the old building. Brian Taylor moved into a simple office in his factory in Manningtree in Essex. Bernard Robinson occupied a room at Tallent that was lined with rows of Coke and Seven-Up cans.

The new bosses worked long hours, and they expected their managers to work hard too – without a lengthy hiatus half way through the day. It was fun for company visitors to play a new game, called Austerity: deduct points for a slap-up lunch in a splendid dining room; add them for stale and curling sandwiches with a cup of tea; double points for nothing at all. It was the 1970s status game with the rules reversed.

The rules were often broken with cars. There were tycoons who drove around in Fords or Volvos, but quite a few Austerity players treated themselves to really nice motors when they could. Brian Taylor bought himself a Bentley Turbo R, one of the most expensive cars in the world, and the standard executive fleet machine for even the most serious player was the far-from-modest Jaguar. It was nothing to do with status, but a lot to do with the fascination so many British men have with the motor car.

The emergence of the predators could be explained by the combination of the recession and the bull market, which provided them with their opportunities. But there was more to it than that: they were early fruits of the seeds of "enterprise culture" that the Tories had sown.

When Sir Keith Joseph complained that there were not enough entrepreneurs to create the wealth Britain needed, he was hoping that a thousand tiny enterprises would spring up, and eventually carpet the country with humming new factories. That, despite a vast range of initiatives, had yet to happen.

A reason was that there were much easier ways of making money than by setting up a machine in a tiny factory and trying to grow from there; one of them was to buy a massive great factory, and to make it smaller – standard predatory technique. To that extent, a new entrepreneurial spirit had indeed been created, but had then shot off in an unintended direction. Parallels could be drawn with Anthony Barber and his boom, or Frankenstein and his monster. The question was, would these new monsters prove a force for good or evil?

The predators were an interesting bunch. They had all the different motivations that have ever driven self-made men. Nigel Rudd came from a poor family; so did Brian Taylor. John Ashcroft was fascinated by dealmaking. In some cases, it was difficult to identify the driver. Harry Solomon gives a disarmingly simple explanation for joining up with David Thompson to form Hillsdown: "I wasn't a very good solicitor, and I always wanted to earn some capital. There was no way I was ever going to build up capital before." His motivation, he said (when his personal shareholding was worth £27 million), was his fear of insecurity. "That is one of the reasons I keep on doing it."

The parallel with old-fashioned entrepreneurs was particularly strong in the sense that all the predatory companies were led by one person, or maybe two. They were closer to the original Austin, Morris or Alfred Herbert than to the teams that ran most companies. The founders formed the dynamo within the organisation and some, like Nigel Rudd, openly admitted that the company would best be broken up after they had departed.

If the predators should ever come to trial, accused of the corruption of British manufacturing, their defence would rest on their effectiveness at changing the people inside industry. Not just by sacking one lot and putting in another; but by liberating suppressed talent that could have stayed suppressed for ever.

It was an article of faith with the predators that poorly run companies invariably had masses of capable, frustrated managers who had been stopped from rising to the top by more plausible, but often less able people who accelerated past them. The predators became particularly adept at picking out those with unexploited talent and thrusting them into positions of responsibility. Their salaries would invariably be linked closely to performance, with bonuses that could kick earnings to glamorous City-like levels: often they would be given a "share of the action", a massive allocation of stocks or stock options.

Harry Solomon's belief in the power of the middle manager was unbounded. "Many of the people who work with us and are successful, are exactly the same people who worked for failing companies," he says. "I believe that people can go through their working lives without fulfilling their potential – most people want to give much more than they are allowed to. If you give them confidence, normally they will succeed." Solomon and Thompson learnt that from their first big deal, when they bought Lockwoods. "We found that whilst the directors were not very good, underneath there was an enormous reservoir of talent – people who knew exactly what was wrong with the business: grey people who at first sight seemed unprepossessing and who come over worst are often those who have most to offer." By contrast, "those who seem to come over well, but are not saying anything, are very dangerous".

Although Hillsdown's takeovers were agreed, Solomon would not hesitate to remove managers if he thought they were over-interested in status or had no rapport with the staff. Sometimes, it was clear enough that there needed to be changes. "If they had three dining rooms, we knew they had the wrong team," says Solomon. At other times it was more difficult – and here he had to cope with a fundamental problem in British industrial society. "There were people who could play the game bloody well," he says. "Many of them shone at meetings – they were well educated and superficially very good; but they would tell you what you wanted to hear. It would be unfair to say that they were chinless wonders, but some of them felt that they were

## People Upheaval

better than other people, when they were not."

(Interestingly, another sort of outsider, the Frenchman Henri Debuisser who became personnel director of Rank Xerox in the early 1970s, sounds the same note: "There's always a bit of a problem with the British education system. You still have a number of people who fool you. They seem extremely bright and they talk very well but they don't produce any output. The trouble with public schools is that they encourage an admiration for people who express themselves well without it mattering what they say – it's the packaging that counts.")

Predators tended to have similar strategies for bashing their new acquisitions into shape. Picking new managers was one part; the others were installing their own control procedures and, crucially, changing the culture or feel of the company. These all tended to be done in one go, following a set "post-acquisition" strategy.

Williams Holdings' agent of change was called the Post Acquisition Management Team or, more usually, the Hit Squad. As soon as a company was taken over, its senior managers were given a pep talk, and left a questionnaire. A typical question was: "If you had just inherited this business, what ten things would you do to improve it?" Within a week, the Hit Squad arrived in its black BMWs with tinted windows. The aim was to cause shock. "People are left in no doubt that the management has changed," says Rudd. "It is not good enough to saunter in at 10am on a Monday and announce you are the new owners."

The Hit Squad consisted of a team of top-notch managers – specialists in accounts, marketing, production, capital control, whatever. It worked alongside the existing management, trying to find out which businesses were making or losing money and why, whether increased capital expenditure would help, and how the management needed to be shaken up. Apart from solving these problems, the aim was to integrate the company into Williams: "A business will never really be part of the group culture if you leave it alone," Rudd believed. "You need to get in there, shake it around, sometimes remove managers who are not up to the mark." The main aim, he said, was to remove uncertainty. If there were to be cuts or management changes, they were announced immediately.

At Coloroll, John Ashcroft instituted a similar régime, with the accent, predictably, on analysis. There were three stages to his post-acquisition routine – assessment, reconstruction and rehabilitation. Assessment could take four weeks: in the first week, a "management audit checklist" was produced by a team of reporting accountants. "There's usually a busload of them parked on the hard shoulder," says Ashcroft. This was a comprehensive analysis of the business over the previous five years, with margins, costs and so on analysed to the last detail. Then, each manager in the company was asked to give a presentation on his or her department, how it fitted in with other departments, and how they foresaw life under Coloroll. One of the main aims was to identify which managers had the best grip on their work.

When taking over publicly quoted companies, as Coloroll or Williams did, it was often difficult to know exactly what they had bought. It was, says Ashcroft, like playing cheat at cards. "You never know what's in there until you turn the hand over."

Hillsdown did not make hostile takeovers. Its logic was in buying companies that could see the benefit of being part of the group, so there was not the need for shock that Williams or Coloroll had to induce. Hillsdown had a less formal acquisition procedure than Williams or Coloroll, but the aims were the same. "We tend to go down ourselves and ask people what the business should be doing. Sometimes we ask them to put in writing what they would do if they had bought the company," Solomon explains.

He was more cautious than the other predators: sometimes Hillsdown acted very fast, cutting dead wood out immediately. At other times, it decided to wait. Because the bids were uncontested, it could see all the cards before it picked them up and was in a better position to make such decisions.

It was comparatively simple for predators to impose change on the companies they had taken over. But what of the companies that had struggled through the recession and had emerged, with new or partly renewed top management but with strategies, products and cultures unchanged? They had to change themselves from within; and that was altogether a trickier matter.

# 15

# *Shake-up*

In the worst-ravaged sectors, the new managers were like architects in a post-war wilderness. They could build what they wanted, and what they built now would decide the shape of the industrial landscape for years to come. If they made the right decisions, British industry would survive and even prosper; if they did not, it would wander back on to a perilous path. For the first time for many years, they really were in charge of the destinies of their companies: the government would not interfere, nor would trade unions. In the private sector at least, strikes had almost evaporated.

The shock of the recession had been far greater than the government had intended. That had one overwhelming advantage: managers could make changes far more radical than they could have contemplated before. It was a unique opportunity, and some businessmen were already taking advantage of it.

Exactly how a company should be organised had already been the subject of many millions of words and much expensive advice. That British Leyland had been moulded into a gigantic monolith was not the result of plain stupidity on the part of the people in charge. It was, after all, the way that Ford was successfully run. Neither, when Michael Edwardes decided to chop the group up into bits, because he thought it was more important to give individual operations their identity back, was he following a universal truth. For such decisions were a matter of individual judgment, often influenced by fad.

Business fads, usually propagated by the management consultants who sponsored them, arose because business was a very imprecise science. In the 1960s and 1970s, McKinsey, the American grandfather of consultants, had persuaded large companies all over the world to install a complex structure called matrix management. A McKinsey director then stood up and said it was not such a good idea, and everyone simplified their structures to the minimum. Owen Green tells a story of how he met ICI's Maurice Hodgson in the 1960s: " 'We've just been McKinseyed,' he said. I'd never heard the name then so I asked what that meant. He said that the structures have all got to be brought together under tight control: it's all quite interesting." Fifteen years later, Green was talking to Hodgson again. "I asked how it was going. 'Oh,' he said, 'we've just been McKinseyed – we're decentralising the whole thing again.' "

Then there was the debate about whether companies should diversify, to stop themselves becoming over-reliant on one industry, or concentrate their resources on what they did best. In the 1960s and 1970s, diversification had been in vogue but in 1982, a book by the American McKinsey consultants Tom Peters and Robert Waterman implored managers to "stick to their knitting". As the book sold 5 million copies, and became a best seller in the USA, many clearly thought their advice worth taking. Although it would back conglomerates if they were aggressive enough, the City broadly supported this view. Owen Green was among those who did not. "Focus," he says, "is hocus pocus."

After the recession, though, it was fairly clear which companies had survived best: those that combined tight financial control with a high level of autonomy for managers. They were given more responsibility, and directly rewarded or blamed for success or failure (often through generous share option schemes): the spread of computers made the monitoring process easier. A vertically integrated structure may have been sensible in the 1950s when shortages meant that it was important to control suppliers: now, a less cumbersome "flat" arrangement – with lots of units reporting directly to the centre – seemed more efficient. These were BTR, Hanson, GEC techniques, so it made sense for other companies to try them too.

John Harvey-Jones applied a modified version to ICI, and it worked. Within two years of his taking over as chairman on April Fool's Day, 1982, the group had made a £900 million pre-tax profit and was being held up as a shining light for the rest of industry to follow. Perhaps here, after all, was a management fad that approached an absolute truth?

The fact that Harvey-Jones was in charge at all was a sign that everyone in ICI accepted the need for change. The public came to know and like him because, with his long hair and bright ties, he seemed most unlike a standard businessman; he was very nearly glamorous. But his peers admired him too: they liked his straightforward common sense combined with an ability to cut through to the heart of a problem. But he was, as he said himself, a high-risk choice, because he had already declared that he was prepared to tamper with ICI's venerable systems. He believed not just in change, but in change for its own sake. "I'm more interested in speed than direction," he said. "Once you get moving, you can sort of veer and tack. But the important thing is, you're moving."[1]

ICI's pre-Harvey-Jones similarities to the civil service were extensive. Its headquarters bureaucracy was massive – 1,000 people beavering away along two miles of oak-panelled corridor in Imperial Chemical House, across the Thames from Lambeth Palace. It was packed full of bright young planners, some of whom must have spotted the imminent catastrophe of 1980. But if they did, the committee-ridden structure meant that the great vessel could not change course in time.

The nautical metaphor became irresistible when Harvey-Jones took over. As a small boy, he had lived in a palace in India where his father was a Maharajah's adviser. He had been packed off to prep school in England, and then went via Dartmouth Naval College into the Navy. After the war, he had become an intelligence officer, learning Russian and German, and doing a stint in the Cabinet Office. But in 1956, when he was a Lieutenant Commander, his daughter contracted polio; he decided to leave and take a shore-based job as a works study officer for ICI on Teesside. He rose rapidly and by 1980 was a deputy chairman, and therefore eligible to succeed Sir Maurice Hodgson.

The process of selection, or rather election, of an ICI chairman seemed to owe more to the Vatican than to the navy or civil service. After two overwhelmingly powerful chairmen, the autocratic Lord Magowan (who formed ICI in 1926 and was only ousted in 1950, at the age of 75), and Sir Paul Chambers, an ambitious financier who ruled from 1960 to 1968, ICI had installed its new democratic system. The chairman, first among equals, was elected by vote of all the candidates, and had a limited term of office – in Harvey-Jones' case, five years. It was a unique process that allowed the unorthodox likes of Harvey-Jones to leap into the hot seat, without giving him too long to destroy the organisation if he was going to.

The tough but straightforward part – slashing a third of the British workforce, closing factories, and withdrawing from complete areas of operation – had been carried out under Hodgson. It was already paying off: in 1982, the group managed a £300 million profit, although high-margin divisions like pharmaceuticals and agrochemicals were buoying up the loss-making petrochemicals, plastics, fibres, and dyes businesses.

Harvey-Jones considered turning the board into a holding company – the BTR model. He believed the current nostrum: that if you chopped a company up into small enough bits, gave people responsibility, and kept an eye on them through numbers, they would do the job for you. But he decided that efficiency could be lost through oversimplification: the potential cross-benefits of letting the subsidiaries work with each other were too great to throw away. The current command structure had to be changed somehow, though: it made running the divisions easy, changing them almost impossible. The board would establish "councils" and "policy groups" by the dozen, but action would rarely follow: with directors responsible for particular businesses, interest groups and mini-lobbies would build up to stop radical action. For example, ICI had tried to move away from bulk commodity chemicals in the 1970s, but found after ten years that they took up the same proportion of production as before, because they had grown through their own momentum. There was, Harvey-Jones thought, a lack of overall control.

What he did was to tell the board it should stick to its knitting – it should restrict its role to doing what it alone could do: planning the future shape of the group, and keeping an eye on individual divisions. The executive board was slimmed from twelve to eight, and policy groups were swept away. Head office staff was halved and responsibility for the individual businesses was pushed on to the shoulders of Principal Executive Officers, who were in charge of the eleven new international businesses. Each was run, virtually, as a separate company, although it would be given a "strategic thrust" by the board – to act as a cash cow or to expand into Japan, for example. A board director would act as a "link man" for each business and territory, but would not be expected to lobby on their behalf. Harvey-Jones coined a new word for this – "non-advocacy".

By splitting the group up into businesses, the board could make sure that one did not sap another's strength. The fast-growing pharmaceutical division kept up its research and development spending even as the organics and fibres divisions were being cut savagely back.

The process by which businesses were monitored was similar to BTR's. Once annual budgets had been agreed, the board would check profit and cash figures every quarter. Only if they deviated from the targets set would it involve itself; otherwise, chief executives would be allowed to spend their allocated funds as they chose. The budgets would be decided in a concentrated annual session, known as "hell fortnight", when the board would go through each division's plan with its chief executive. The businesses would work on rolling three-year plans, although the board tried to keep part of its multifocals focused on the future – a team of futurologists was employed to stare way into the next century. It was a mix of long-range planning and short-term thinking, and it seemed to work well.

Then, ICI turned its attention from structure to strategy. Like other companies, it no longer had to ask, "What are we here for?" It knew the answer now: to make a profit. The moral agonising over job creation or social responsibility had, thanks to the City and government, been hurled mercilessly out of the window.

There were three main ways of increasing profit in a manufacturing operation. Sales could be increased, costs could be cut, or profit as a percentage of sales – profitability – could be boosted. Or a combination of any of these.

As British managers looked round the world, and dipped into writings on the subject, it became clear that the most successful country, Japan, favoured the sales-led approach. As Japanese motorbike makers had demonstrated in the 1970s, big enough volumes would lead to lower overheads per unit produced, and thus to lower costs.

There were two rational reasons and one irrational one why this approach was universally avoided in Britain. First, the Japanese built up their original volume within an internally competitive, but externally protected, market. That meant that companies would be both large and highly efficient by the time they unleashed themselves on the outside world. That was impossible in the open British market. Second, any bid to go for volume would inevitably mean sacrificing profits in the short to medium run. Stories of Toyota operating in the USA for ten years before making a profit were enough to make any British manager wince with helplessness. The problem was one of finance. Japanese firms could tap the very low rates of the post office savings system (the Japanese were prepared to invest at a *negative* interest rate). The British, like the Americans, had to cope with short-termism, either from the City or from the banks. Any manager who told his bankers that he would be making losses for a few years while he built up his market share would, more likely than not, be deemed unfit to run a business.

Third, and irrationally, post-recession managers shied away from growth. Bigness had led them to where they were now; they must never stumble into that trap again.

So they fell back on the second and third options. Slowly, they were learning the secret of cutting costs, mainly from the Japanese. And they were working out what every economist learnt at school – that the best way of increasing profitability was to operate in something less than a perfectly free market: the nearer they could get to a monopoly, the higher the price they could charge. It had little to do with the free market, but it made business sense.

One of the favourite routes to monopoly was by digging out a niche. There was nothing new about the idea of a "niche product". Wedgwood china, Guinness, the better mousetrap – they were all products that had something different about them (originally a genuine distinction, perhaps now just a name) that meant that they did not have to compete just on price. Sir Owen Green and Lord Weinstock had long encouraged their managers to favour specialist, high margin products. They had a fundamental tenet: "profit is sanity, volume is vanity." Now other firms, with the encouragement of the City, were following the same line.

The "niche" was something that had to be dug out of a market, but the size of the market could vary. Manufacturers of chemicals would seek out "global" ones: the world was their market place. Other products could establish themselves most naturally in a region, a country, or a group of countries. A niche product that owed its power to a brand name would obviously be effective only in the area where it was promoted.

For ICI the move to global niches was an obvious reaction to the damage the non-specialist operations had suffered in 1980: the fibres operation had lost £86 million on sales of only £450 million, petrochemicals had lost £44 million, organic chemicals and plastics £35 million apiece. The classic niche division, patent-protected pharmaceuticals, had made £66 million on sales of £346 million. Resources were poured into this and into other specialist divisions, while dozens of projects that could create niches in the future were set in motion – in biotechnology, for instance. "If we have twenty businesses, and eighteen are doing OK, we can afford some seed corn businesses," says current chairman Sir Denys Henderson. "Hopefully, the chairman in 2010 will say those guys back in the 1980s were quite brave – they kept plant breeding on." This was an advantage of sheer size; if a company was rich enough, it could get away with a little judicious long-termism.

A niche has to be defined by what customers want: it is, after all, a niche in the market. Under Harvey-Jones, the non-specialist divisions started to concentrate their considerably reduced resources on looking for niches – in other words, on becoming more specialist. Whereas in the past their research scientists tended to produce something then look for a customer, now the process was reversed. The chairman called it "inventing to order". In the petrochemicals division, researchers tried to create the best sort of film for video, while the fibres laboratories concentrated on a new cotton lookalike.

At Blakely, Tony Rodgers and his team decided to steer the Organics division away from its traditional dependence on textile dyes, and towards high-growth areas such as colour reprographics and electronics. Even as plants were shut and consolidated, the managers started a careful investment programme, halving the workforce and product range, but increasing the total size of the business. Throughout the process, they worked closely with the unions, sharing confidential information with them. "It was never once abused," says Rodgers.

He continued to put great emphasis on keeping up morale. In 1983, when rumours were flying about the closure of the Blakely site, Rodgers ordered 1,000 trees – "Blakely Forest" – to be planted. "I was accused of littering the place with trees when we were still in the throes of a redundancy programme," he says. "But semi-symbolic changes reinforced substantive change." When the division finally broke even, in 1987, he sent all 6,000 staff a bottle of champagne. "I got a note saying the last time that had happened was when the Queen got married."

Meanwhile, Harvey-Jones had decided that the group must be further internationalised. Having been born of a cartel that kept it within the bounds of the Empire, ICI had found itself trapped after the war in low-growth markets, and with a heavy reliance on the UK (which by the 1980s absorbed only 4 per cent of the world's chemicals). In the early 1960s, it had begun to build plants and make acquisitions in Europe, and ten years later had started to develop in the USA. American sales were $250 million in 1972 and $1.2 billion in 1983 – but Harvey-Jones wanted to boost them further. To spot acquisitions, a group called the A-Team was set up, and Henderson was put in charge of it.

Beatrice Foods was undergoing Wall Street's latest torture – a leveraged buy-out. The LBO was a particularly obnoxious creature of the bull market, which owed much more to asset stripping than it did to the creation of industrial efficiency. What happened was that an "investor", or group of investors, would borrow vast amounts of money from the Wall Street banks, buy a company, then break it up and sell off as much as they could. Because the deal was so highly geared, or leveraged, the question of investing in the company was rarely considered.

But ICI at least benefited from the LBO when it heard that Beatrice's chemicals divisions was for sale. The A-Team was shifted to New York by Concorde within twenty-four hours, and $750 million later, ICI found itself with a new group that was particularly strong in advanced plastics for the aerospace industry; US sales rose to $2 billion a year.

Beatrice was the first big foreign acquisition. Over the next two years, a series of takeovers outside the UK reshaped the group. In 1980, the UK had accounted for 42 per cent of sales, the Americas for 16 per cent and continental Europe for 19 per cent. By 1987, the figures were: Britain 25, the Americas 27 and continental Europe 25 per cent. While overall sales rose from £5.7 billion to £11.1 billion, British sales crept up from £2.4 billion to £2.7 billion. The shift in employment patterns was also striking. In 1980, ICI employed 143,200 people, of whom 84,300 were in the UK. In 1987, total employment was 127,800, with 55,800 in the UK: the British share fell from 59 per cent to 44 per cent.

Reinforcing the internationalisation, Harvey-Jones brought in foreign non-executive directors – Shoichi Saba, president of Toshiba, Walther Kiep, managing director of a large German insurance company and Tom Wyman, chairman of CBS (later, Paul Volcker, chairman of the US Federal Reserve was also drafted in). Not surprisingly, the chairman was accused by one British union leader of being "the human face at the head of the dole queue". But ICI epitomised one of the main trends of the 1980s. In its 1980 annual report, it provided a detailed breakdown of exports from the UK – they were £1,173 billion, more than any other industrial company's. In the 1987 report, no figure was given, while much play was made of the manufacturing operations overseas. Even though the statutory obligation to declare the export figure had disappeared in the early 1980s, the tone of the report reflected the shift that took place in many companies as part of their restructuring. Exporting from the UK, which had been regarded as an absolute good, bringing wealth and employment most directly into the country, became unfashionable. Companies said they needed to invest abroad to get "closer to their markets" and, of course, to dig out global niches.

The Labour Party was not impressed; it wanted to force companies to repatriate their capital and to reinvest it in Britain. And, critics pointed out, the Japanese had grown so wealthy entirely on the back of direct exports – why couldn't we?

There probably was an element of faddism in the moves to multinationalisation by British companies, and they did indisputably reduce the size of the UK's manufacturing base. But they were also inevitable, arising as they did from a combination of the abolition of exchange controls in 1979 and a lot of hard navel-staring by post-recessional British companies. A company that stared harder than most, before moving overseas and nichewards, was Lucas.

Lucas had had a bad press. In the USA, it was known as the Prince of Darkness. "Why do the British drink warm beer?" they used to ask. "Because their fridges are powered by Lucas batteries." The group had struggled through the recession on the back of the aerospace activities it had built up in the 1970s but, unlike ICI, it had not then slipped back into profit. After losing 10,000 workers in 1980/81, it was repeatedly hit by collapsing markets and niggling industrial problems. Its expansion into the USA had been useful in 1980, but two years later helped to push it back into the red. The group continued to shed jobs, closing an American plant in 1982, and making 1,200 electrical workers redundant in January 1983.

Lucas was looking vulnerable, if not to the receiver, then to predators. Much of its valuable cash was being poured into redundancy payments and although it kept up its research and development spending, it had a profile the City did not like: too close to the motor sector, based in the Midlands, altogether too much like Dunlop. It also seemed uncertain which way it was going, and harboured rather too many managers who hung on to the old complacency. Not surprisingly, the financial pages buzzed with takeover rumours.

Chairman Godfrey Messervy, together with Gill and manufacturing director John Parnaby (who had come from Dunlop), developed a scheme to reshape the company. Individual businesses were given more autonomy, but were then told that they would have to identify their competitors, see how they matched up, and produce "Competitive Achievement Plans" showing how they could bridge any gap. Parnaby, an expert in Japanese production systems, helped businesses draw up their CAPs: they had to produce targets that were both financial and physical – including measures of quality, lead times, value-added, and so on. Each plan would be sternly examined by Messervy, Gill and Parnaby. If it was practicable, it would be approved. If not, the business would be sold or closed down. It was a neat way of catching up with foreign rivals while pushing responsibility down to the business managers – and shaking any remaining complacency out of the system.

The CAPs also gave an "objective" evaluation of the way the company should be moving, and arrived unsurprisingly at high technology global niches. Of the 150 business units that had to produce CAPs, thirty-five failed the test, and were discarded. The company was completely reshaped, as the strong aerospace divisions were bolstered by acquisitions, mainly in the USA, while the basic car parts divisions were discarded. In 1979, UK sales were £606 million, overseas sales £547 million; in 1987, the figures were £696 million and £1,125 million. Aerospace sales tripled: their share of the total rose from 11 to 24 per cent, while the share of the automotive divisions dropped from 80 per cent to 67 per cent. The changes paid off: profits rose steadily, and breached £100 million in 1987.

Behind these statistics was an ever-increasing concentration on the areas in which Lucas had achieved what was known as "global mass", where its sales were high enough to generate funds that could consolidate its lead. It was reaping the rewards of keeping its R&D spending up, even during the worst of the recession. On the vehicle side, it had reached global mass in diesel injection, petrol engine management and anti-lock braking systems: it was even challenging the mighty Bosch on its home ground of Germany. In aerospace, it had dug out several niches, including engine controls and fin systems on guided missiles.

In 1987, it finally had to bow out of the car lighting market – with which its name used to be synonymous – selling that business to Fiat. It was like London knocking down St Paul's, but it finally started to convince the cynics in the City that Lucas should perhaps not be written off.

A niche could be based on genuine differentiation – Lucas's sales were based on superior technology, ICI's on specialist products. But it was also possible to create a niche purely through marketing. Retailers had long understood that, as had companies like Coloroll, which divided the market up into any number of different segments, and produced a range of wallpaper for each. Peter Ward saved Rolls-Royce Motor Cars by creating a niche on a grander scale: he called it the Bentley.

Born in Stafford in 1945, Ward was brought up in Coventry with motor oil in his blood; his father was an engineer and sometime racing mechanic for Alvis. He was toughened up in the sales organisation of British Leyland during the troubled 1970s, and rose rapidly to become sales director of its parts subsidiary, Unipart. He transferred to

Peugeot Talbot, to take charge of its parts division and, towards the end of 1982, got a phone call from Rolls-Royce; would he, they asked, like to set up a marketing division for them? "I wondered if they had got the right Peter Ward," he remembers. But they had: Sir David Plastow, the ex-Rolls-Royce chief who was now running the parent group Vickers, had recognised that the car company's catastrophic sales plunge had been largely due to a lack of marketing skill. What crusty old Rolls needed, he thought, was an assault by a new-style manager. No one had ever thought much about selling Rolls-Royces before: they were supposed to sell themselves.

So, at the age of 37 (and looking 27), the amiable Ward was given the task of saving one of the few products that had had a global niche for seventy years. When he arrived, the company was having to discount its cars, labour relations were troubled, and profits had almost disappeared. Ward was a professional marketing man, and started applying the same techniques to Rollers as he had to boxes of bits. He was shocked by the resistance to change, but realised that he too came as something of a shock to the traditionalists at Crewe. "I talked about gross margins and price elasticity," he says. "All the things that make people think you have two heads."

His first move was to start advertising second-hand Rolls-Royces. Everyone thought he was mad – until they realised that first it could create brand loyalty among younger drivers, and second that if the price of old cars was pushed up, owners were more likely to trade them in for new ones. The ploy worked (and the technique has since been copied by most of the other luxury car makers). Then he looked around for a new niche.

Bentley had been one of the great names of motor racing in the 1920s, winning the Le Mans twenty-four-hour-race five times in seven years. But, under Rolls-Royce ownership from 1931, it had lost almost all its identity. For many years, Bentleys were just Rolls-Royces with different radiators; in 1980, they accounted for only 5 per cent of total sales. Ward decided that a wonderful name was going to waste. It could be used, he thought, to dig out another niche without compromising the existing one, just as a new soap powder could be used to capture a different segment of the market.

He went for a younger customer than the typical Rolls owner – "somebody who grew up on hairy British sports cars and Beach Boys music and who could no longer get granny in the back of the Porsche". He tempted them by launching a "down market" model, the Bentley Eight, in 1984. At a mere £48,000, £6,000 less than the cheapest Rolls-Royce, it was within range of the most expensive Mercedes or Porsche buyers – just. It had a mesh grill reminiscent of the old racing Bentleys, but was otherwise pretty much a Rolls-Royce. Having thus brought the marque into the minds of younger people, he leapfrogged the Rolls-Royce range and the next year produced the Bentley Turbo R. It cost £85,000 and was very, very fast. With straightline acceleration better than most Ferraris, and roadholding that belied its two-ton weight, some of the motoring press thought that for the first time since Edwardian times, the company really did deserve the "Best Car in the World" tag it had so arrogantly hung on to. In 1986, just to ram the new image home, a Turbo R won sixteen national speed and endurance records: never mind that nobody else had bothered to compete for some of them since the 1920s, they all helped the image.

Ward had created a completely new market without damaging his old one. Rolls-Royces would still be bought by the staider company chairman; Bentleys would go to people who would otherwise have bought Aston Martins, Porsches or Ferraris. It was a brilliant piece of niche creation, and it paid off. In 1987, a third of all the cars the company sold were Bentleys, and Rolls-Royce made a profit of £22 million.

Although the British had always been rather good at marketing consumer products – having picked up and taken wholeheartedly to American techniques in the 1950s and 1960s – the great concentration of power in the hands of the supermarkets had left many manufacturers behind. It was these that Hillsdown "rescued", gathering them up into a fold that could both talk on equal terms to the likes of Sainsbury and Tesco, and service their needs better. In the two years after it floated in February 1985, Solomon and Thompson had bought forty-two companies, mostly food producers, for a total of £220 million.

Hillsdown became particularly good at raising profitability. Convenience foods had been around for a long time, but in the 1980s they started to move upmarket; fish fingers were no longer considered good enough for sophisticated British consumers – they must now be offered "doigts de poisson" for which, with a bit of parsley balanced delicately on top, they would be asked to pay three times as much. It was, depending on your point of view, something of a con-trick, or clever niche-creation. Hillsdown took the latter view, and encouraged its companies to "add value" in this way wherever they could. The products would be sold either under one of Hillsdown's brand names (including Lockwoods, Smedley's, Daylay and FMC), or as a supermarket own brand. Although by the end of 1986, Hillsdown was the biggest British producer of poultry, eggs and meat, and the largest canner, its name did not appear on a single product.

Part of Imperial Group's poultry division, which Hillsdown bought in 1984, was Buxted, the chicken company. It had lost £14 million the year before. The Hillsdown people quickly discovered that the previous manager – who came from Imperial's Golden Wonder crisp subsidiary – had been trying to sell chickens in the same way, taking on a big sales force and trying to create a "brand image". It did not work, and the advertising costs dragged the company deep into the red. Solomon and Thompson picked out David Newton, a manager from another part of the Imperial purchase, and let him get to work. He chopped the sales force, cut the advertising budget, and revamped the strategy. Instead of plain frozen chickens, he started producing fresh ones, most of them treated in some way to provide added value. Adding a pellet of garlic butter to a chicken portion did not cost much – but it turned the piece into chicken kiev which could then be sold, highly profitably, to a supermarket chain.

In November 1985, Nigel Rudd and Brian McGowan of Williams Holdings bought a company that set their thoughts moving in a new direction. They paid Burmah Oil £8 million for the British operations of Rawlplug, well-known to DIY enthusiasts, and even better known to builders for its fixing devices. In 1984, it had made £1.5 million on a £20 million turnover. A little bit of shaking, combined with £2 million in investment, pushed the profit up to £5 million within two years. More important than that, Rawlplug made them realise the usefulness of brand names as a way of protecting a niche: another firm could make a device exactly like a Rawlplug, but unless it was called a Rawlplug, it could not sell for the same price. All that the owner of a brand name had to do was to keep it fresh with copious advertising.

Rawlplug taught them two other things. First, that the building products business was a good one to be in. The low tech engineering sector had, more or less, been "cleaned up" by the predators. And even those companies that were still badly managed had become more highly rated by the City, their share prices pulled up by the expectation of a predator's swoop. The housing market, on the other hand, was starting to boom: over the next three years, it was to ripple out from the South-East to take in even the most devastated regions; and the closely related DIY market was having its own revolution on the back of the "supermarkets" like Texas and B&Q. So Rudd and McGowan followed the boom: for the next few years, most of their acquisitions would be in housebuilding or DIY. Williams became rather less of a conglomerate, and a little more like Hillsdown or Coloroll. By 1987, by which time Williams was turning over half a billion pounds, 60 per cent of the group's profits came from these areas.

Second, Rawlplug hammered home the point that its ability to earn "above normal" profits was enhanced by the fact that it was the dominant supplier in the fixings market. If another company produced a similar product, also branded and backed up by marketing, prices and therefore profit margins would have to be cut back. From now on, Williams looked for companies that were either the market leader or a good second (sometimes this was preferable because it would not attract the combined firepower of the other makers). In this way, they could ensure that either they controlled the prices within the sector, or were at least not at the mercy of price-cutting by the leaders. Ideally, this market power would be backed up by having the lowest production costs in the sector. "If there was a real drop in demand, we would buy the opposition," says Rudd, "and if it went on dropping, we would be the ones who put the lights out, because of our strategy of being the lowest cost producer."

The combined strategy of having a high market share and low costs was reminiscent of the Japanese high volume approach. But Williams (and the scores of other British firms that were following the same path) achieved their aims without needing to increase overall volumes. Instead, they bought market share by taking over other companies and when they invested, it was always to cut costs, never to expand output. Efficiency increased; the size of British industry did not.

The fact that the predators invested at all set them apart from the asset strippers. Some were better at investing than others: a company taken over by BTR was, for instance, more likely to be given new equipment than a Hanson acquisition. Hillsdown and BTR both made a point of investing as soon as they could to let the workforce know that they were not going to run a factory down. "We have always tried to find a reason for spending on a piece of equipment," says Green, "preferably big and showy and as soon as possible after we have acquired a company."

When Williams bought Rawlplug, it discovered that its main product, the Rawlbolt, was being cast rather than pressed. After investing in a pressing machine, the cost of each item was brought down by 30 per cent. Sometimes, Rudd says, managers would have to be bullied into investing. "I was told about a project where if we mixed some material on site, we would save £1 million a year. The machine to do that would cost £1.5 million, so I asked why it had not been done. The answer came back that they had had rather a lot of capital expenditure in the last year, and they felt we would not be willing to sanction any more. They had had years of people telling them not to spend any money: it was a mental thing."

That was a rather simple example of how investment could cut costs. But the process by which other companies had been streamlining themselves, leading to what some people called a productivity miracle, was anything but simple. It had been going on since the recession, and had started with a realisation that if a company wanted to survive in this ever-internationalising world, it would have to be as efficient as anyone, anywhere. And that usually meant in Japan.

By 1980, the history of the photocopier industry was following a well-trodden path – Western technology, Western development, Western sloppiness: markets lost to super-efficient Japanese. As Rank Xerox's profits collapsed from £277 million in 1977 (when it was the ninth most profitable company in Britain) to £166 million in 1983, the fortunes of its chief Japanese competitor were going the other way. From 1975 to 1985, Canon's sales grew eightfold and profit twenty-two times. But in 1980, Rank Xerox decided to strike back. It sent teams out around the photocopier world to seek out the best levels of efficiency in a range of production areas; these included the cost of the product, the rate of defects, the size of inventory and the level of overheads. They noted the best standards down, and called them benchmarks. Most, not surprisingly, were found in Japan.

By 1988, chief executive Roland Magnin declared, Rank Xerox must meet or beat the benchmarks. That would not be easy. In 1980, there was one defect in every machine it made; the benchmark was two per 1,000 machines. In 1980, 17 per cent of components were rejected; the benchmark was 0.5 per cent. In 1980 RX was holding three months of inventory; the benchmark was two weeks. And so on.

They were tough targets, but there was one obvious way to meet them – to copy the people who set them. All eyes turned east, towards Japan: it was time to copy the copiers.

# 16

# The engineers fight back

Among Britain's complex pattern of social divisions one, little recognised in the past, had been eating away at the country's industrial fabric: the divide between engineers and everyone else. If industry as a whole was beyond the Establishment pale, people who actually made things were somewhere out on the horizon.

As a result, they were an aggressively proud lot. They loved the cacophony and sweet oily smells of their factories, and they despised non-engineers who tried to tell them what they could or could not do in them. Until the expansion of graduate courses in the 1960s, formal engineering training was limited. On the job experience combined with rugged common sense was, it was reckoned, a match for anything that could be taught in a classroom. As the vast majority of British manufacturing companies were built up by engineers, they tended to be run in their image: they were rugged, and they made rugged products. Modern cost controls and fancy marketing were unknown and unused. Massive machines made in Birmingham will be thumping away in the depths of Africa long after their Japanese replacements in Britain have fallen to bits. But the company in Birmingham will no longer exist – because the Japanese machines did what the customer wanted better, and more cheaply.

In 1980, Sir Monty Finniston, the Scottish engineer and former chairman of British Steel, produced a report demanding that something be done to upgrade both the quality and status of engineers. The government ignored most of his proposals, which would have cost money, but did allow the creation of the Engineering Council, an umbrella training organisation for the eighty or so jealously independent engineering institutes. Slowly, training courses and exams started to expand.

The potential weakness of companies dominated by engineers had been demonstrated by the collapse of Rolls-Royce in 1971. The directors tended to be so obsessed with the finer points of engineering that they did not keep an overall view of what was happening to costs or markets. In Germany, engineers worked closely with all the other disciplines. In Japan, the aim was to produce renaissance managers, with a good grasp of all the disciplines and a clear view of the whole.

As the Germans and Japanese had struggled to recover from the war, they had naturally to use both astute marketing and careful cost control to win markets (often from the British). The marketing problem was now being tackled in the UK – but what of costs? In particular, why were the Japanese able to make things so much more efficiently and to such high quality? Why were the gaps found by Rank Xerox in its benchmark-seeking expedition quite so gaping?

There are three basic forms of production: process, mass and batch. Process production, of liquids such as chemicals or paint, is inherently the most efficient; by definition, it is based on constant flow. Here, efficiency is gained by making the plant as large as possible, as capital costs do not vary greatly with capacity. Mass production, pioneered by Henry Ford, aims to apply the same flow technique to discrete products: the product moves along a conveyor belt and is gradually built up by workers who perform the same task on each unit. However, it is expensive to set up, and only works where there are long runs with limited variation between products: cars are the obvious example. Most factories have batch production. Here, a product is converted from one state to another by passing through a series of lathes, presses or other general machine tools; it may be welded or assembled as well. After a certain sized batch has been run off, the machines have to be set up to process the next product.

As factories grew, so did sloppiness – especially in batch production. There was only one rule when adding a new machine: put it near its brothers; that looked neat, and allowed the skilled operators to stay together. Otherwise, growth was higgledy piggledy. A new building might be tacked on the back, a new type of machine jammed in wherever it would fit. The philosophy, if there was one, was to have more than you needed, just in case. There should be more machines, more workers, more stores than were strictly necessary. Batch runs would be as long as possible, to cut the number of "set-ups". The result was an excess of everything, giving infinite potential for confusion.

Ingersoll Engineers, a manufacturing consultancy, painted this picture of chaos in a 1982 report. Most elements, it claimed, were still to be found in 80 to 90 per cent of British metalworking plants. "Forklift trucks hauling materials everywhere, damaging and losing parts. Progress chasers trying to find parts for the next operation. Operators and machines waiting for the chaser to find the parts. Aisles full of people dodging forklift trucks. Inspectors trying to separate bad parts from good throughout the process. Large piles of partly finished products, occupying floor space, being lost or damaged. And multiple layers of supervision trying to control the whole affair."[1]

British companies were particularly bad at controlling the variety of products they made. They did not usually try: metalbashers had a tradition of being able to provide anything their customers wanted, and their managers were proud of that. In one (now defunct) British company in the 1970s, sales of £5 million were achieved with more than 100 different products; in its main German rival, a £13 million turnover was generated by twenty products. That made each British part more expensive: the more set-ups between products, the less time the machines were running, the greater the cost per unit.

The basis of the Japanese "laser beam attacks" was the narrowness of their product ranges. In the late 1960s, Japanese bearing manufacturers had lines with between a quarter and a half the variety of their Western rivals: the volumes they could generate were produced at costs that caused havoc in the West, and very nearly drove the Swedish giant SKF to the wall. The British motorcycle industry was, of course, a victim of the same strategy.

British managers had failed to spot that they should be restricting their product ranges; they had allowed their factories to become cluttered, inefficient and overmanned. Overmanning was particularly bad among "indirect labour": the store keepers, maintenance men, quality controllers, clerks, and all the other employees who were not operating machines that made things. In a typical British factory, the ratio of direct to indirect workers was two to three; in a Japanese factory, it was two to one.

The advantage of being so bad was that it was relatively easy to get better fast. That was what happened in the immediate aftermath of the recession, when the drop in output was quickly recovered with a workforce 20 per cent smaller. Indirect workers (including managers) took the brunt of redundancies; flexible working agreements forced greater efficiency on those who were left. But the next step was trickier.

## The Engineers Fight Back

Productivity had improved but, by every measure, it was still way behind the best Continental levels and not within sight of the Japanese.

By the time the House of Lords Committee reported, at the end of 1985, there was plenty of evidence that great changes were in preparation, even if few had yet been implemented. As ICI had proved, companies could be restructured rapidly, new marketing strategies could be drawn up in a matter of months, subsidiaries could be bought and sold in days. But the process of streamlining production was a long, slow and tedious one: it involved massive reorganisation and, above all, completely changing the way that managers and operators worked. Most companies were barely out of hospital: it was unreasonable to expect them to be sprinting yet.

The most advanced manufacturing plants in Britain were American-owned. Some like IBM, had been using Japanese techniques before the Japanese had themselves. Other American companies, seeing Japanese plants springing up in America and Japanese cars driving their own off the roads at home, had reacted with typical vigour. Encouraged by squadrons of management consultants, they were way ahead of the Europeans in meeting "le défi japonais", and were now transplanting their newly learned philosophies across the Atlantic.

Not far behind the Americans was a rather surprising group – the British state-owned giants. In the 1970s they had started to look at ways to survive in the long term. Invariably, their plans had included massive investment which their political masters, determined to avoid job losses, had approved. British Steel, Rolls-Royce and British Leyland had thus become remarkably well equipped. Because they were not inhibited by the financial constraints of a quoted British company, nationalised manufacturers were able to behave rather more like their Japanese or German rivals.

Like many of the more successful modern engineers, Andy Barr was a tough and articulate Scot. As head of manufacturing at Austin Rover he, alone among British Leyland's senior managers, had survived all the purges and ructions. If anyone's personality was stamped on the volume car division, it was his. As the recession died down, his factories did not seem in bad shape. Manpower had been drastically reduced, and the changes forced through by Michael Edwardes – an insistence on flexible working and the abolition of almost all demarcation – had had an immediate effect on efficiency. The lines were now running virtually without interruption.

But Barr knew he could not stop there, because Austin Rover had a unique problem: it was not big enough to conduct a slogging match with the big six European manufacturers (each of which produced four times more cars than it did), neither did it have the marketing power to move up to, say, the BMW level. For the moment at least, as a small, non-specialist producer, it was stuck with the worst of both worlds – high costs and low prices. But it was a problem that, Barr concluded, could be bypassed by the judicious use of technology.

Even though the Metro line, which started up in 1980, was at the time the most modern in Europe, it was its younger cousin at Cowley, which made Maestros and Montegos, that showed how Barr really intended to confound the laws of economics. There, one production line was producing four different models, the Maestro, the Maestro van, the Montego and the Montego estate. Robots would dart forward, make a weld on, say, a Maestro, then wait for the next car to come along. If this was a Montego, it would make the appropriate weld for that; its programme would tell it what to do. In a conventional factory, Cowley's unit costs could be achieved only if one model was being made on the line: in effect, the economies of scale were being quadrupled.

The Japanese were the masters of flexible manufacturing systems such as these, and Barr picked and chose from their armoury of techniques. He was in a better position than most to do so. Austin Rover's rebadged Hondas – the Acclaim and the 200 – had already saved its bacon and throughout the early 1980s, the two companies' engineers worked together to produce the first genuinely joint product, which would become the Rover 800. In common with the other car makers, Austin Rover adopted the Japanese approach to subcontractors, trading longer contracts for better quality. The company classified its suppliers according to quality: in 1981, 30 per cent were grade A; in 1985, 90 per cent were.

As Barr was able to put more reliance on his suppliers, he could make another step towards Japan. A fundamental difference between Western and Japanese car companies was that while Ford and Austin Rover made as much of a car as they could themselves, Toyota and Nissan made as little as possible. They believed in concentrating on what they were good at, which meant assembling and painting cars: everything that could be was subcontracted out. This was sticking to the knitting, Japanese style.

The next stage of Barr's technical strategy allowed him to move along this route. His aim was to run Austin Rover's factories on the most advanced CAD/CAM – computer aided design/computer aided manufacture – system in the world. CAD was already established: its ability to produce 3-D images on screen had confirmed its superiority over the drawing board. But ARG then started to push it back into its suppliers: it would design a car on a screen but give only rough guidance about, say, the headlamps. That image would be passed down the line to the headlamp manufacturer, who would use his CAD screen to design the part in detail: software houses sprang up all over the West Midlands to help iron out the bugs in a complex system.

In theory, once the car had been designed, the computer could be used to direct the machines that made the dies and tools needed to manufacture each component. Finally, and this would be CAM, or rather CIM (computer integrated manufacture), the central brain would drive the whole manufacturing system. A change could be made to the design on the CAD system, the new tooling would automatically be created, and the new part would quickly be put into production. Such a system would need a lot of clever people, so Barr set about raising the "intellectual level" of the company. In the 1970s, the group had employed 200 graduates; by 1985, with half the number of employees, it had 700. The intake of 100 engineering graduates a year was sent straight out again to Warwick University. ARG endowed a £5 million unit there, where problems such as emission controls could be studied in depth. Other graduates were put through a part-time masters course. The aim, Barr said, was "to produce the car company with the best brains". He hoped they would be a match for the brawn of the other Europeans.

The truck factory at Leyland, which opened in the midst of the 1980 recession, was another showpiece of technology. The commercial vehicle division had been starved of investment after the creation of British Leyland and by 1975 was producing a large number of products from thirteen small and inefficient factories. When better and cheaper lorries started arriving from Scandinavia, Germany and Holland, Edwardes launched a programme that combined a £350 million replacement of the Leyland range with a series of closures; production would be concentrated at the new factory, which cost £33 million.

At first sight, the plant did not seem sophisticated. Painted in cheerful primary colours, it moved great chassis round on a slowly moving track; at each station, a worker would tighten a set of bolts or fix a new part. The factory looked like a giant Meccano set – comfortingly old-fashioned compared with the glitzy robotics of the car lines. Yet it was reckoned to be the most advanced truck plant in Europe when it opened. It had a huge stores area, with two and a half pigeon holes allowed for each part, and a fleet of forklift trucks to move them around. Seven minicomputers linked to an IBM mainframe controlled every tiny detail of the production process. In 1982, the director of manufacturing, John Gilchrist, came across a system in America that he thought would help the plant (which had been dogged by technical teething troubles) work more efficiently. It was called MRP2, standing for Manufacturing Resources Planning, and its purpose was to help control the flow of parts. MRP1, which originated in the USA in the 1960s, would take the factory's production schedule, based on sales figures and forecasts, work out what parts should be needed when, and automatically order them to be delivered at the right time. MRP2 took this a stage further, taking account of the numbers of machines, people and tools that were available. It could even stop the track if it discovered the parts were not available.

MRP's parts control showed how far a powerful computer system could be used to bring order into a factory and, in the process, cut stocks as far as possible. As a load of components arrived at the Leyland plant, an operator at the gatehouse would key in their details; they might be directed to the lineside, more likely to the stores. There a forklift truck driver would pick them up and store them in the right slot; as he did so, he would tell the computer by keying in the number of the batch. When a part was moved on to the line, its number would be keyed in again, along with the vehicle it was destined for. Finally, as the lorry chassis was completed, another number would be entered – and all its constituent parts would be discounted by the system.

To work properly, the MRP computer had to be fed with vast amounts of information, and it had to be correct: if anyone short-circuited the system, for example by taking a part and not entering it, it would soon become inaccurate. The Oliver Wight Organisation, which developed MRP, reckoned that most British companies were getting 10 per cent or less of its potential benefits. Leyland became the only British-owned company in the UK to be awarded the Class A status by Wight.

These British Leyland plants used what might be called the American, as opposed to the Japanese, solution to productivity growth, based on maximum use of technology. Machines were used to replace people – welders, designers, stock-keepers – but there were no basic changes in the way production was organised. That was fine for companies that had the resources to do it properly, but too many managers decided that a little bit of automation would provide a quick fix for their long-term problems. It rarely did.

The government encouraged them in this belief. Computers were now affordable by the smallest company and, it was widely believed, Japanese success was largely due to their new robots; better still, robots did not strike. In the aftermath of the recession, the story peddled by sellers of robots (for welding) or Computer Numerically Controlled lathes (which could be programmed to carry out a sequence of operations) was a convincing one: they offered better flexibility, accuracy and quality.

But most companies that paid up for this equipment soon discovered that they were painting their houses without shoring up the walls first. Ingersoll Engineers carried out several dozen government-funded studies to see how companies could best automate. In a quarter, it recommended installation of new equipment. In 30 per cent, it told companies to turn themselves upside down before installing anything. And in the rest, it concluded that all was needed was a thorough reorganisation. "Panacea technology", Ingersoll told its surprised clients, was not the answer to their problems.

Companies that used machines to replace people were not attacking costs where they were greatest. In a typical metalworking plant, direct labour accounted for about 10 per cent of manufacturing costs, overhead costs (including indirect labour) for 35 per cent, while the remaining 55 per cent was taken up by materials. British companies, obsessed with the productivity of labour, tended to hack away at the 10 per cent. That was not sensible, particularly in a relatively low-wage economy.

What was really needed was common sense combined with discipline. Fortunately the engineering sector was benefiting more than any other from the great expansion in technical universities of the 1960s. It had brought in a new wave of highly educated engineers many of whom, moreover, had worked in the more advanced atmosphere of American multinationals. They understood the need to catch up with the best companies in the world.

Rolls-Royce, the aero engine company and founder inmate of the state's lame duckery, showed how common sense and automation should blend together. In 1978, it sent a team of managers and workers' representatives to America to find out whether it matched up to its competitors. It discovered it did not, by a frightening margin. American companies were working at double the Derby factory's machining rate, they had half the number of inspectors, lower stock levels, and a much more flexible workforce.

Three years later, it threw itself into a programme to up-end the way it produced compressor and turbine discs, some of the most expensive items in a jet engine. Any reduction in their six-month manufacturing time would swiftly feed through to profits. Ingersoll Engineers, which helped develop the AIMS – Advanced Integrated Manufacturing System – project, came up with several ideas. Most were based on simple common sense. One, to use a different sort of forging, cut the amount of metal that had to be bought by 45 per cent: this immediately saved £1 million a year. By rearranging the discs into families with common characteristics, the same tooling could be used for more than one disc: in 1978, disc manufacturing involved 2,000 different turning tools – that was cut to 100.

Where automation was used, it sprang out of a need to plan production more efficiently. Much of the £4 million was spent on moving existing equipment around, rather than installing brand new machinery. But the new factory looked bang up-to-date, mainly because it used that icon of progress, the driverless truck. Automated Guided Vehicles were installed for one good reason. Whereas previously a forklift would carry ten discs, and production would therefore have to work in batches of at least 10, the AGV carried only one, so that discs would never be sitting around. If they were not in machines, they would be in transit. With each disc worth up to £40,000, that was important. Fitting this ultra sophisticated AGV system into a Victorian factory tested the ingenuity of the engineers: three air raid shelters and any number of girders had to be coped with. One of the shelters was turned into a coolant reservoir for a machine tool. The workforce also had to be told, as part of their training, that there was no greater sin than leaving metal foil from cigarette packets on the floor: it could trick an AGV into unloading its valuable burden on to the floor.

The Japanese were great people for sitting around in groups trying to solve problems; that was the basis of their quality circles. Rolls-Royce applied the same brainstorming technique while AIMS was being developed. For example, it previously took seventy-two hours to reset the machines used to cut slots in the rim of the discs. The existing operators and engineers sat down with a handful of bright young graduates. The graduates came up with new ideas, some feasible, some fanciful, and the setting time was whittled down to one and a half hours. When AIMS started working in July 1985, the disc production time was down from six months to six weeks. The reduction in inventory holdings alone saved £4.6 million – more than the entire cost of the project.

The AIMS project proved that extraordinary improvements could be made by the ruthless application of logic. What Ingersoll was doing at Rolls-Royce was to eliminate waste – wasted time, wasted material, wasted people. That, as Ingersoll knew quite well, was one of the basic tenets of Japanese manufacturing.

*

## The Engineers Fight Back

Soon after he had become chief executive of Tallent Engineering in 1980, Bernard Robinson headed off to the Far East to find out why Japanese products were so much better than British ones. He wanted to buy Japanese machine tools, and asked their manufacturers to arrange for him to see them working locally. That was how he came to visit his Japanese equivalents, and to discover some of their secrets.

Just after the war, two American academics called Edwards Deming and Joseph Juran touted their ideas round the mighty American industrial machine, and were rebuffed by the proud "Fordist" production men. So they went to Japan, where pride was a scarce commodity, and found managers leaping at their ideas like starving children. The Deming award for quality, which was founded in 1951, became the ultimate goal for all Japanese companies. Deming and Juran were celebrities, their names known by the man on the Yokohama omnibus, and nothing baffled the Japanese more than the discovery that they were virtually unknown in their own country.

The philosophy was based on a simple idea: that if no mistakes were made in the first place, no time need be wasted on rectifications, production schedules could be planned with certainty and much of the paraphernalia of stocks and inspection could be abolished. Whereas a British company would typically regard quality as adding cost – through armies of inspectors standing at the end of a line – Deming taught that it did the opposite: it could make those inspectors redundant. It also led to higher quality: if a product was made correctly in the first place, rather than being bashed into shape afterwards (a time-honoured engineering practice), it had a higher chance of working properly. It could be summed up in an elderly cliché: prevention is better than cure.

As first-time quality improved, the "cushions" could be removed from the manufacturing system. Traditional belt-and-braces – an insistence on extra stocks, extra manning, extra machinery – was replaced by the flimsiest of tights that would rip apart at the slightest disturbance. Overheads fell as stocks shrank and floor space was freed up; indirect workers in the stock room and on forklift trucks became redundant.

To achieve the goal of getting everything "right first time", Deming came up with the concept of Total Quality Control. Everyone, he said, had a customer: for the marketing person that was clear enough; but the production line worker had a customer too – the next man on the line, who would have to cope with his mistake; the clerk in accounts had to fill out forms clearly, because someone had to read them; and even the telephone operator had to satisfy the callers. The idea was that by pushing responsibility down the line, the entire workforce would be busy trying to solve problems, rather than just the 10 per cent at the top.

Quality was simply defined as what would satisfy the customer, rather than adherence to some set standard in the British fashion. Indeed, the concept of "standards" was inimical to Japanese thinking: they believed in constant improvement, with no plateau ever reached. This management by a thousand changes was known as "Kaizen".

Total quality was particularly appealing to the Japanese because it required the business to be looked at as a whole, rather than chopping it up into distinct functional chunks (marketing, finance, quality control, etc.) as the Westerners were wont to do: Japanese managers were first managers, second specialists. That was why, when a Japanese product was designed, the whole process of making and selling it would be taken into consideration. In an old-style Western factory, a product would be designed, and the engineers would be told to make it. In Japan, part of the design function would be to make it easy to assemble: that was one of the reasons why their products were more reliable.

A whole raft of techniques was built up around the total quality principle. Many were alien to British traditions and, Bernard Robinson realised, could only be introduced slowly and with relentless training. Furthermore, the swashbuckling tradition of British engineering had to be jettisoned and replaced by the rigour of the Japanese. Everything must be planned, and there must be no deviation from that plan. It meant that a great British strength – the ability to improvise – would have to be deliberately suppressed. In many circumstances that would be a disaster; in a factory, it could work. All it needed was discipline.

Robinson's first Japanese move was painless enough. If responsibility for quality was to be pushed right down to the operator, he had to be given the wherewithal to control that quality. The most tangible tool was Statistical Process Control, a device for keeping machines in true: Tallent introduced it in 1980. Five times an hour, the operator would take a component from his machine, and measure it in each direction. Each measurement would be transferred to a piece of graph paper, where a line would develop through the day, showing how far the machine was straying from its true setting. If the line showed a tendency to stray beyond control limits, the machine would be recalibrated, and the reason for its straying investigated. Statistical analysis of these data meant that the performance of the machine could be monitored and, with adjustment, the control limits could be brought closer together. In theory, no unacceptable parts could be produced, and the need for quality inspectors would be abolished. Robinson did not trust the theory: he kept his inspectors.

When Tallent won its long-term contract to supply Sierra suspension arms to Ford, Robinson had installed £3.5 million worth of equipment. At the same time, he took the opportunity of introducing one of the most fundamental of Japanese techniques – cell working. This was an excuse to take the serried ranks of identical machines, jumble them up and create little villages of different machines. Each village was a mini-factory, responsible for making a particular group of products and manned by between three and fifteen people.

The implications of this system were sweeping. The most tangible advantage was that the widget under construction, instead of travelling long-haul between distant banks of machines – with all the forklift trucks or transport systems that that involved – would make a brief and pleasant journey round the tiny cell area. It was less likely to get lost or damaged en route, and would not have to be pursued by progress chasers. But more fundamental to the cell idea was the team spirit it was supposed to create. Workers who had been alienated by monotonous production lines would suddenly find themselves part of a group and doing more varied work. Cells would be expected not only to check their own quality, but also to measure their input and output and solve their own problems; the idea was that they would feel a close identity with, even pride for, the product of their mini-factory.

Robinson found that the new arrangements, which gave the workers more varied jobs, were welcomed. But there were worries in other firms. Unions were suspicious of cells because they undermined the whole concept of crafts from which they had drawn their strength: in a cell, each member would be expected to be capable of every job. And some managers did not like them either. If problems were going to be solved by the operators themselves, what was their role? It was a good question, because the system did indeed need fewer levels of management. The flattening of company structures was continuing right down to the shopfloor level.

In 1985, Tallent introduced "just-in-time" or Kanban, the most difficult to implement and therefore extreme form of Japaneseness. It was an attempt to manufacture discrete products in the same way as, say, chemicals: raw materials in one end, finished product out the other, with no interruptions and no blockages. The idea was that parts should be "pulled" by demand from the next stage, rather than "pushed" according to some pre-arranged schedule. At Toyota, where the name Kanban came from, it had been perfected. Little trolleys, each designed to take a certain number of a certain part, would trundle around the factory floor delivering parts from one machine to the next. But, and here was the just-in-time bit, the first machine would operate only if there was a trolley waiting to be filled up: it would never grind away building up a stack of unused parts next to it, and the parts it made would always arrive "just-in-time" at the next machine. In fact, the machines would hardly ever have to stop, because Toyota's engineers would adjust their speed so that there should always be trolleys waiting; but Kanban could cope with changes in demand of about 10 per cent.

JIT worked best if it was extended throughout the production process, with the raw materials arriving just-in-time at a component maker, the components arriving just-in-time at the main factory, and moving just-in-time from one machine to the next before passing on to the wholesaler or retailer. Pushed to its ultimate extent by Toyota, the process went right through to the final customer: he would go into his showroom in downtown Osaka and order a particular model with particular features. The order would be keyed direct into the factory's scheduling and a just-in-time system would start pulling all the parts together to present him with a complete car at the showroom – just two days later. Nothing would ever be hanging around waiting for something to be done to it – it was perfectly efficient.

Just-in-time had an effect on two levels. By reducing stores to an absolute minimum (in extreme cases, no more than two hours' supplies would be kept), and abolishing "work-in-progress" between machines, it both saved on financing costs and provided a simpler, uncluttered factory: things were less likely to get lost, and it was easy to see which parts were available, where. But more than that, it was the ultimate enforcer of "right first time". If anything went wrong – a delivery did not arrive on time, a part was substandard – there was an almost immediate result: the line would grind to a halt. In that way, it was the opposite of MRP: MRP was designed to control complexity; JIT did not allow it to exist. If anything went wrong at Toyota, there was only one thing the workers could do: shout "*Jidoka*!" or "Stop the line!"

Bernard Robinson thought Tallent was running smoothly, with minimum inventories. Then, when he installed JIT, there was chaos. "It showed up a hell of a lot of weaknesses," he says, "because previously we had had enough stock to cover five- or ten-minute breakdowns. In the short run, it caused us more losses than it saved because we had not totally grasped that JIT was inseparable from quality." He was not alone – the one blip in Japan's headlong productivity growth was when companies en masse were introducing JIT in the 1970s.

In the West, managers looked mainly at just-in-time in terms of delivery of supplies: they would still be scheduled, but much more frequently. It was easy to fake JIT: if a client insisted on it, his supplier could produce as he always had, put the parts into store, and deliver them according to the JIT schedule. In other words, the stock was being shifted from buyer to seller – robbing Peter to pay Paul.

This was inevitable unless every company in the production chain was prepared to offer JIT. Tallent did deliver just-in-time to his customers, all big companies, and its steel was delivered up to eight times a day. From Robinson's point of view, he was achieving full just-in-time. But there was a catch: his steel supplier was a stockholder, which dribbled steel in as he needed it. It was faking just-in-time. American and Japanese steel companies were already offering JIT deliveries to all their customers: British Steel, for all its new efficiency, was not. Until it did, just-in-time could never work properly in the engineering industry.

To ease these changes in, Robinson started to behave like a Japanese manager. He became obsessive about training, and sent his workers on course after course to extract their maximum potential. He started to wander around the factory as much as he could – to see, to be seen, to ask questions, to be asked questions. It was called "management by walking around". And he introduced regular briefing sessions for the workers, stopping the factory for half an hour every month to brief them on progress. These were all rather obvious things to do – but they were still rare enough in British factories.

Robinson avoided quality circles, which were attracting praise and opprobrium in varied measure. As the first Japanese technique to hit the public consciousness, they came to be a symbol of this new alien order, and were treated with due suspicion. By the end of 1982, there were about 100 British companies with quality circles: Marks & Spencer was encouraging its suppliers to introduce them, and the experience of some companies, like Jaguar, was excellent. Workers would be trained to approach problems in a systematic way, and would carry out their own investigations before presenting the management with solutions.

Ford had had to give quality circles up when the unions withdrew their cooperation after only a year: they claimed that they were a way both of bypassing themselves, and of getting something for nothing out of the workers. It seemed that the company had pushed the scheme through without properly consulting the unions or lower management – both of whom felt threatened. Jaguar's Egan, on the other hand, was able to unite the workforce in the face of desperate financial difficulties and a common enemy: British Leyland.

By the middle of the 1980s, British managers had become almost as hungry for knowledge of these new systems as the Japanese had been thirty years earlier. They trotted along to seminars in their hundreds, and lapped up the writings of the gurus. By far the biggest impetus for the new techniques came from the consultants. Companies such as the Boston Consulting Group, PA and Ingersoll Engineers took Japanese ideas, wrapped them in friendly Western packaging and set about hawking them to British Industry. They did not find them difficult to sell.

By 1985, there were few results from the Japanese experiment, but those that there were looked promising. State-financed Jaguar had turned itself around (admittedly from an appalling base) by concentrating on quality – complete with quality circles and long-term supplier relationships. Rank Xerox had combined a fierce round of cuts with full-blown Japanesification. Just-in-time had been introduced, the number of suppliers slashed and the remainder given longer contracts; a "leadership through quality" (total quality) programme was introduced. Everywhere in its plants, slogans exhorting workers to ever higher quality appeared on the walls: local children were encouraged to paint pictures on the theme of "quality". The benchmark programme allowed Rank Xerox's progress to be accurately monitored. Inventories were down from ninety-nine days of stock in 1980 to forty-five days in 1984 (the benchmark was seventeen). The ratio of indirect to direct workers had dropped from 1.4 to 0.5. In 1980, 30,000 parts out of every million that Rank Xerox bought from its suppliers were defective; by 1984, that was down to 3,150 (the benchmark was 950). And the defects per hundred machines made fell from ninety-one to twenty-one: the aim was two. The results were encouraging. As Rolls-Royce had shown with AIMS, it was possible to achieve "step changes" – that is, improvements so large that they put the company on a different level of performance. The more conventional measure of profits confirmed Rank Xerox's recovery, but did not tell half the story: they bottomed out in 1983 at £166 million, and then started to rise gently. By 1985, they stood at £200 million.

But these were exceptions. The mass of British industry had to emulate Rank Xerox, Rolls-Royce or Jaguar if they were to catch up with the world and, in 1985, that seemed a distant prospect. Tallent had SPC, a handful of cells and the beginnings of just-in-time. As far as Robinson was concerned, the Japanese experiment was only just beginning – yet he was far ahead of the majority of engineering companies. Even the giant Lucas, under the guidance of manufacturing guru John Parnaby, was only just starting to install Japanese systems. Everyone's thinking was moving in the same direction; few people were actually doing anything.

Nissan's plant in Washington, Tyne and Wear, was due to start operating in 1986. That would be the real test for the likes of Tallent and Lucas. Would Nissan accept their quality, or not? Lucas in particular had cause to worry. Its changes might stand it in good stead in the long run, but this was the middle of the 1980s, the peak of the bull run – and a time when the short run was getting ever shorter. Takeover mania was about to undergo its own step change.

# 17

# *Big bids and politics*

Nineteen eighty-six was designated Industry Year by the government. The idea of the first industrial country needing such a thing seemed faintly ridiculous – but who could mock any attempt at turning people's minds towards factories? It was a fairly hopeless cause, though, for 1986 turned out to be the Year of the City. October's Big Bang, with the strange salaries that went with it, transfixed the attention of school leavers and new graduates, reducing even further industry's chances of attracting the best talent.

But, precisely because of the growing influence of the City, 1986 did indeed turn out to be a year of great change for industry. A tremor of takeover activity, a powerful aftershock to the recession, rolled through the economy and reshaped complete industrial sectors – and some of the largest companies in Britain. Behind this new activity lay the ever closer relationship between industrialists and their financiers, a relationship that, in the Guinness affair, was finally shown to be a little too cosy.

As City and industry were drawn together by ever larger takeovers, so government came to join the fun. The Westland affair and the government's attempts to boot British Leyland out of its lame duckery added an element of Whitehall farce to the whole affair.

In the autumn of 1985, the merger wagon gained new momentum. The City, bulging with machismo, let it be known that no bid was too big for it, and merchant bankers urged their more aggressive industrial clients to think enormous: the way to get credibility was to buy something big, and previously gentle companies started sharpening their claws. The easy pickings were no longer among the engineering smokestacks, where the weak and the lame had already been picked clean, but in the relatively unscathed land of food and drink. The recession had caused the odd shiver and reorganisation but, until the threat of takeover arrived, there was no real pressure for change. That, inevitably, meant there had been no real change.

As Hillsdown had already demonstrated in one narrow sector, there were good commercial reasons why a certain amount of merging was necessary: the growing power of the big retailers was one, the gradual trend towards internationalisation and "global brands" was another. Many of the takeover bids were backed by commercial logic but, with the bull market racing ahead, inevitably the balance sheet wizards leapt in on the act. The stakes in the City game became so high, and techniques so sophisticated, that it was certain that sooner or later, someone would be caught cheating.

The scene was set by Guinness' takeover of Bell's, the whisky company, in August 1985. Guinness had one of the best-known "niche products" in the world but, for all its clever advertising, it had been suffering from the rise of lighter beers. It bubbled with cash, which it had not been very good at spending; diversifications into mushroom farming, pharmaceuticals and film finance had gone wrong, so the Guinness family looked for someone to inject some fizz into it.

On October 1, 1981, Ernest Saunders was appointed managing director. He had everything it took to be a new-style manager. He was an outsider – his parents were refugees from Vienna (he said he never felt fully part of English society) – and when he left Cambridge in 1960, he did not follow his contemporaries on the standard path to law or the civil service, but joined the American company 3M as a graduate trainee. He moved on to J Walter Thomson, the advertising giant, and then to Beecham: he was fascinated by marketing, where American techniques of product positioning, brand image and advertising were starting to become much more sophisticated. Self-confident and able, he rose rapidly and by 1975 was international marketing manager of Nestlé in Switzerland. There, he made a name countering accusations that Nestlé's powdered milk fostered malnutrition in the third world.

Once at Guinness, Saunders rebooted the advertising campaign by sacking his old employer J Walter Thomson, and set about unburdening the group of its excess baggage – 150 companies making products that ranged from sweets to light fittings. In the best British tradition, Guinness had no proper accounting systems which meant, he said, that he was "flying through fog". He brought in advisers to help him bring some order to the company, most notably Olivier Roux, a bright young Frenchman from the management consultancy Bain & Co; he became, in effect, the finance director.

In the summer of 1984, he started buying; his acquisitions seemed at first sight as irrational as the companies he had shed, but they were all based on Saunders' shrewd judgment of what the market wanted: Champney's health farms, the British franchise for 7-Eleven all night supermarkets, and a ragbag of other stores. He was moving Guinness away from its old base of making things to one of selling things. In August 1985, he bought Bell's after a tussle with the management. That moved Guinness back towards manufacturing but, more importantly, marked the start of the great food and drink takeover spree.

At the beginning of September, Allied-Lyons – cakes, ice cream, beer – found itself being eyed up for the chopping board by a new sort of predator. John Elliott, a tough ex-McKinsey man from Melbourne, had built up a powerful Australian empire called Elders IXL: it was best known for its Foster's lager. Now, along with the likes of Alan Bond and Robert Holmes à Court, he had come foraging in Europe. Elliott finally made a formal bid for Allied-Lyons on October 22. It was for £1.8 billion – nearly twice the size of the previous British record bid (BAT's for Eagle Star) – and it marked the arrival in Britain of the leveraged buy-out. Allied-Lyons was four times the size of Elders IXL, and Elliott could buy it only if he then broke it up to repay the banks: he would keep only the brewing group for himself. There was nothing unusual about the idea of a break-up – it was a Hanson speciality – but the level of debt set this bid apart.

It would be a deal on the American model. If British takeovers had tended to oil the wheels of industrial restructuring, on Wall Street the bull market had just led to asset-stripping on an ever grander scale. The likes of Ivan Boesky, T Boone Pickens and the Eton-educated Sir James Goldsmith were financial geniuses with no interest in management. They would seek out companies whose assets they considered undervalued by the stock market, buy a big chunk of equity, and see what happened. Sometimes, they would end up owning the company through a leveraged buy-out financed by bank debt and high interest "junk bonds". Because most of the equity, the shareholders funds that acted as a financial cushion, was turned into debt, both the risks and rewards were magnified. If the business went well, the small slug of equity held by the financiers would increase rapidly in value; but if it went badly, the tiny cushion would soon disappear. That led either to deeply conservative management, with costs squeezed wherever possible, or to the break-up of the company to pay off the debt.

## BIG BIDS AND POLITICS

At other times, the threatened company would escape, but only by buying the predator's now inflated shares off him. It was called greenmail, and it caused havoc. Banks, fed up with the third world, were only too eager to back such deals.

No one suggested that Elliott was a greenmailer, but his deal had a strong whiff of Wall Street about it. On Thursday, December 5, the secretary of state for trade and industry, Leon Brittan, agreed that the bid should be referred to the Monopolies Commission because of worries about the financing arrangements. The government was, for the moment at least, putting an import ban on highly leveraged deals.

But already that week, the attention of the business world had wandered from the Elders/Allied fight. Three deals, together worth more than £4 billion, had been announced. GEC had made its £1.2 billion bid for Plessey on Tuesday, but the day before there had been two even bigger bids. James Gulliver, who ran the Scottish-based Argyll foods, offered £1.8 billion for Distillers, the biggest whisky company in the world and – in the view of many – one of the most sloppily managed companies in Britain. That bid had been expected for several months and it was the other announcement that caused the real stir. Imperial Group, the sixteenth biggest industrial company in Britain and second in the consumer side only to Unilever, said it was going to pay £1.22 billion for United Biscuits in an agreed merger.

There was something funny about this deal, for the buying company was undoubtedly the less dynamic of the two. UB, although less than half the size of Imperial, was run by one of the more effective consumer businessmen, the Edinburgh-based Sir Hector Laing. Very much one of Mrs Thatcher's "us", he had ramped up UB's productivity to stand up to the growing power of the supermarket chains. Imperial, on the other hand, was a lumbering giant that had grown great on its domination of the home tobacco market. In the 1960s, it had diversified, picking up Courage and dozens of food and drink companies. It had improved much since Geoffrey Kent took over in 1981 – when profits had hit a nadir of £5 million – but, as Hillsdown had proved by turning round its poultry businesses with great ease, it was still riddled with inefficiency. Imps had also been financially weakened by a disastrous acquisition in 1980 – the US restaurant and hotel chain Howard Johnson – which had now been sold.

The combined group's £6 billion sales would include, from Imperial, Courage beer and John Player cigarettes, and from UB, McVitie's biscuits, Terry's chocolates and KP snacks. The cash puffed out by Imperial's slowly declining but massive tobacco interests could be easily inhaled by UB, which needed to invest heavily to win a decent bite of the US cookie market. Kent and Laing both played on the need to create a real international giant in the food industry. It was the same argument that Weinstock was using in his bid for Plessey: that Britain needed companies of a size that could take on the biggest in the world.

The City was unconvinced. It thought the deal was really a defensive "reverse takeover" of Imps by UB, to create a group that could fight off the predators. The predators thought that too.

On Friday, December 6, the day that Jim McAdam was appointed chief executive of the textile giant Coats Patons, the early editions of the papers carried a story that Hanson Trust was about to bid for his company. Hanson already had a 3 to 4 per cent share, so the idea seemed plausible; Coats' shares rose sharply during the day. In the afternoon, McAdam was rung by a man from the *Daily Mail*, telling him that Lord Hanson was having a press conference at 6pm. This, thought McAdam, would be the shortest chief executiveship in history. Then the journalist rang back. "It's not you," he said. "It's Imperial Group."

Hanson had detected that the City was not happy about the UB/Imperial merger, and he needed a victory himself. His attention had swung back towards the UK in the last few years. In 1982, he had paid £95 million for Berec (maker of Ever Ready batteries) and the next year had bought United Drapery Stores for £255 million, then sold most of it off. In 1984, he had bought London Brick for £247 million, increasing its profitability simply by cutting the workforce by 20 per cent and making three price rises in quick succession. But in 1985, his attentions had been repulsed by Powell Duffryn, the engineering group, and Hanson Trust's shares started to lose their gloss.

Hanson's problem was that he needed to buy to grow. His vicelike management always squeezed more profits out of acquisitions but, unlike most of the other predators, he was averse to all but the most conservative capital expenditure. His group's raison d'être was, he admitted, buying and selling companies and the bigger he got, the bigger those companies had to be to generate the same percentage growth. Imperial, with sales of £5 billion, lots of assets and an unloved management, was just the job: he offered £1.8 billion for it.

Kent put up a stout fight: one advertisement headlined "A battery of evidence" claimed that Berec's capital expenditure had been slashed by 50 per cent after being bought, its advanced projects division had been sold, its European operations had been sold off to its main rival, and it had suffered a 20 per cent loss of brand share between 1981 and

1985. Journalists were sent a hamper with a selection of UB and Imps products, followed the next day by an Ever Ready battery and a London brick. The note read: "We did not include this in your hamper yesterday as obviously they do not go together." Merger Mania was indeed a game; the power wielded by financial journalists, who sat at the sidelines awarding each side points, made it seem all the more unreal.

McAdam's reign at Coats Patons did not, it turned out, last very long. Although the textile industry had made an excellent recovery as sterling weakened, there were those within it who were determined to strengthen it against the next downturn – and that meant creating more powerful groups.

Coats Patons was the gentle giant of British textiles. To call it British was something of a misnomer, for its strength had always been making thread abroad. Once the biggest company in the world, a Coats thread factory was to be found in every type of city, most especially in South America. The Scotsman who ran it (it was a Scottish company, and it looked after its own) would be a tough expatriate, whose need to grapple with hyperinflation accounting would be softened by a lifestyle long forgotten in Britain: the group was known as the "rich man's diplomatic service".

In the late 1960s, Coats had moved into UK knitwear and spinning (including Jaeger), following the trend towards vertical integration. As Far Eastern imports started to pour in, that strategy was slammed into reverse, and the second half of the 1970s was spent cutting back its British capacity; in 1970, its Paisley thread mills employed 8,000, in 1980, only 800. Coats' geographical spread saw it through the UK recessions, but also limited the tonic effect of tumbling sterling. Instead, it decided to diversify into a whole range of ventures. It took up eel farming, at first using waste heat from its Paisley mill, then from the Hunterston B power station; and it started to make pacemakers. Both projects failed.

An equally unlikely venture, manufacturing precision zinc components, was a success; its only link with the rest of the group was that it operated in some of the same countries (thirteen of them). Unfortunately for Coats, though, this was not the time to be seen diversifying: "sentiment" had turned against it. That was what the City decided the group should have done. The horror of a Latin American connection (however successful) did not help, and Coats, though profitable and turning over £1 billion by 1984, found itself lumbered with a lowly price/earnings ratio.

McAdam had no illusions about Coats being anything other than a sitting target for predators, but he did want to make sure a link-up was industrially sensible. A genial Scot who had spent most of his working life on the Coats overseas circuit, he would undoubtedly have found Lord Hanson an uncongenial boss.

There was a man, though, who had been weaving the textile sector into a new and tougher fabric, and with whom McAdam had had a series of dinners. He was called David – or more properly Davoud – Alliance, son of a Sephardic Jewish merchant from Teheran. At the age of 16, he had set up as a textile trader, and three years later, in 1951, had arrived in Manchester to look for better quality cotton goods. Lancashire was still the centre of the world cotton industry but, with its family-owned business and old-fashioned machinery, its mills soon came under pressure. Alliance borrowed £8,000 from a moneylender and set about reviving one in Oswaldtwistle that was facing receivership. He kept the mill viable, and after ten years bought a cheap clothing mail order company. By analysing the market carefully, rigorous stock control and careful differentiation to suit market needs, he built it into a major business. Another ten years on, in 1973, he bought his first public company, a bra and corset maker called Spirella. Three years later, he merged it with Vantona, which was already a sizeable textile group.

Alliance broke the group into small units, pushed responsibility down to the managers and built up his credibility with the City; in 1982, it allowed him to buy Carrington Viyella. This company, half owned by ICI, was a grand collection of brand names that had been slaughtered by the recession, losing £85 million in three years. Alliance bought it just as the textile cycle bottomed out, used Vantona's cash to free it from debt, and started to invest in new equipment to provide the flexibility retailers were now demanding. In 1984, Vantona Viyella's turnover was £385 million. In 1985, it took over one of Marks & Spencer's biggest knitwear suppliers, Nottingham Manufacturing: this had £100 million in spare cash which, Alliance thought, could well be spent on Coats Patons.

Unfortunately for Alliance, McAdam had not picked up his discreet signals of interest, and early in January 1986 the Coats man started talking to Ronald Miller, chairman of Dawson International. Dawson was another Scottish company with some thoroughly recession-proof niche products, including Pringle and Ballantyne knitwear. Its sales were a quarter of Coats', but the City liked Miller and gave Dawson a stronger price/earnings ratio. Within eight days, the deal was assembled.

When it was announced on January 27, Alliance was flummoxed.

He immediately called McAdam in Glasgow, who agreed to meet for a chat. Within days, a new deal had been put together: Alliance would pay £711 million for Coats, against the £660 million or so that Dawson had offered. Coats Viyella, the biggest pure textile company in Europe, was created and Alliance was one step closer to creating a properly international group immune from the vagaries of currencies. Ronald Miller was left waiting bitterly at the church door, but made no attempt at reconciliation.

Back in the consumer sector, Ernest Saunders had joined in the battle for Distillers. The directors of the Scottish company, seeing itself drawn ever closer to the jaws of Gulliver, whom they regarded as something of an upstart, looked round for a "white knight" to whisk them away to safety. Their merchant bank advisers at Kleinwort Benson started talking to Guinness, and on Sunday January 19, Saunders and his team spent the day beating out an agreed offer for Distillers. Guinness would offer shares worth £2.34 billion, comfortably topping Argyll's bid. Saunders also agreed to move the merged group's headquarters to Edinburgh, and to appoint a stalwart of the Scottish establishment, the governor of the Bank of Scotland, Sir Thomas Risk, as non-executive chairman. In exchange, and unusually, Distillers would pay Guinness' costs during the bid. Gulliver was not deterred. He came bouncing back on February 5 with a higher bid. Crucially, the offers were in the form of shares: each bidder had to play the old predator's game on a grand scale, urging their share prices up in whatever way they could.

In the midst of all this excitement, another rather different sort of battle was started. It seemed, superficially, to be another takeover fight. A consortium led by Sikorsky of the United States and Fiat of Italy was bidding for Westland, the British helicopter maker that was about to bellyflop into order-less doldrums. Then a rival bidder had appeared, a European consortium headed by British Aerospace and including GEC. Would this be a battle decided by the City? Not a bit of it – it was not much to do with industry at all, but greatly to do with politics.

Early in 1985, it had become clear that Westland would have virtually empty order books after 1986. In 1985, it lost £90 million on a £300 million turnover, and was running out of money fast. In September, Sikorsky started talking to Westland; its chairman, Sir John Cuckney, was receptive. The Secretary of State for Defence, Michael Heseltine, was furious. He was determined to stop Britain's only helicopter maker falling into American hands and, flying directly in the face of Josephism, he used his powerful contacts to assemble a rival consortium. That put him into conflict with Leon Brittan, the trade and industry secretary, who was backing Cuckney and who in turn was backed by Downing Street.

The storm gathered strength around the turn of the year. Soon, Westland and its 11,000 employees were no more than pawns in a byzantine political power struggle. Heseltine walked out of a Cabinet meeting on January 9, and retired to the back benches. On January 24, Brittan resigned too, his reputation smeared by a leak that he had apparently authorised. Little Westland came perilously close to toppling the prime minister herself.

The commercial plot continued to unfold, coloured always by political overtones. The ubiquitous Lord Hanson revealed on January 16 that he had a 15 per cent share in Westland. Whether the holding was purely commercial, or had some political element (Hanson was at the heart of the Thatcher mafia, and was much keener on keeping friends in America than in Europe) was not clear. His holding balanced that of the big helicopter operator, Alan Bristow, who supported the European option. At a meeting to approve the Sikorsky plan, Cuckney could muster only 65 per cent backing, 10 per cent short of what he needed. Immediately there was a hectic round of share buying, with six "mystery shareholders" taking a 20 per cent holding.

Politics affected all big takeovers. One way to beat off a predator was to persuade the Monopolies and Mergers Commission to block the bid; and the power of reference to the MMC (which would itself trigger a six-month hiatus) rested in the hands of the trade and industry secretary. Brittan's successor, Paul Channon, found himself immediately confronted with a decision on whether to refer the UB/Imps arrangement or Hanson/Imps, or both.

On February 12, he cleared Hanson and referred the United Biscuits plan. One broker said that "the illogical bid had been given government blessing and the logical one the cold shoulder". Sir Hector Laing agreed: he called the decision "quite monstrous". The decision had been made, apparently, because the Office of Fair Trading decided that there could well be a monopoly in the crisp and nut market. Ironically, Imps and UB's main argument – that they fitted together so well – was the cause of the referral. Bricks and crisps were, the government thought, a healthier combination. There were those who looked for a dark political side to the move, but with Laing even closer to the prime minister than Hanson, it was difficult to lay a charge of favouritism.

What the referral did show was that the monopolies guidelines were failing to keep up to date with the increasingly international structure of business. They were drawn up along exclusively national lines – and thus, as Lord Weinstock was discovering, could scupper moves that would be good for British industry in the long run.

On the day that Channon made his decisions, February 12, Westland finally fell into the hands of Sikorsky. The casting vote was in the hands of the six unidentified shareholders, hiding behind Swiss bank names. The Westland affair petered out in the same blanket of fog that had covered it throughout.

Hanson's £1.8 billion bid was not high enough to secure Imperial and at 9.30 on February 18, he announced a new offer of £2.32 billion. At precisely the same time, Sir Hector Laing came bouncing back with a straight takeover bid for Imps – for £2.56 billion. It had a rider attached saying that Golden Wonder, part of the snacks business, would be sold off to meet the monopolies objection. By the end of the day, UB had bought 8 per cent of Imperial: it had to raise a £350 million loan to pay for the shares. The Imperial board started lobbying for the UB offer to be accepted.

Meanwhile, the Guinness bid for Distillers had also been referred; the way looked clear for James Gulliver to have his way with the whisky company. But Saunders came back with another bid, this time for £2.35 billion – topping Argyll's offer by £70 million – along with a note promising that it would reduce its share of the UK whisky market to 25 per cent by selling off at least five companies. It was an action replay of UB's ruse.

As if Westland had not created enough political turbulence, the Austin Rover/Land Rover affair came along to stir up a little more. The previous November, the general manager of General Motors' Bedford truck subsidiary, J.T. Battenberg III, had confirmed that he was talking to his opposite number at Leyland Vehicles, Les Wharton, about linking the two companies. These talks were the result of a Wilsonesque move in 1984 by Norman Tebbit, then the trade and industry secretary, to encourage British truck makers to talk to each other to cope with overcapacity. Leyland and Bedford had each lost £60 million in 1984, and the fit was not bad: Leyland, with its modern range and facilities, was strong in the middle of the lorry market, while Bedford was well positioned with its vans – even if they were Japanese designs from GM's associate companies, Isuzu and Suzuki.

British Leyland's subsidiaries were busy taking advantage of their government ownership to force through modernisation programmes. Even as it announced another 750 redundancies in December 1985, Austin Rover was opening its £5 million high technology centre at the

University of Warwick, and Leyland Vehicles said it would spend £9 million on creating a fully Japanesified axle production line at its Albion plant in Glasgow. But, when Austin Rover's 1984 results showed that it had dipped back into loss, pushing its accumulated deficit to almost £1 billion, there were plenty of people who wondered how long the government would tolerate the burden.

On February 1, Roy Hattersley, the Labour deputy leader, revealed (and denounced) a plan to sell the bus and truck operations to General Motors. The government had, in its keenness to offload as much of the group as it could, also offered Land Rover to GM. No sooner had that political hornets' nest been stirred up than Paul Channon confirmed that the government was talking to Ford about selling it Austin Rover. To complete the sale, the bus division would be going to the Laird Group, whose Metro-Cammell subsidiary made buses and trains.

Ironically, if the deals went through, it would be the acquiring companies' workforces that would suffer most. Because BL had been pouring money into both its truck and car operations, its factories would stand a better chance of surviving rationalisation than either the Bedford plant, which GM had long neglected, or Ford's troublesome Dagenham factory. However, it was at BL's West Midlands factories that the strongest opposition emerged: the idea of this heartland of industry becoming a land of branch factories was more than the Midlanders could bear.

As the Opposition, as well as a goodly number of interested Tory backbenchers, demanded that the government rethink, Mrs Thatcher gave an impressive display of ideological dogmatism. She dwelt on the fact that BL had so far absorbed £2 billion of taxpayers' money, and claimed that there was nothing to show for it. She did not mention the fact that British Leyland was now one of the most efficient car producers in the world, or that its current problems owed a great deal to the depressed car market. Austin Rover's managers themselves were worried about quite another matter – the loss of their valuable link with Honda if the company was swallowed by Ford. Ray Horrocks, Austin Rover's chairman, expressed the frustration that managers felt when in mid-March he made a personal statement to the Commons select committee on trade and industry in which he talked of the "cruel and shameful persecution" of British Leyland. He was expressing the complete demoralisation of a company that felt it had done everything it could and might, if left alone and unabused, succeed.

Thanks to the extraordinary post-Westland political climate, the

Ford deal lasted just three days in the public domain before Paul Channon hit it on the head, announcing that talks had been terminated. It turned out that the Cabinet had outvoted the prime minister on the matter.

In the midst of this politically charged climate, other companies came in with opportunistic bids. The engineering company Aveling Barford, as well as respective management teams, wanted both Land Rover and Leyland Bus; Lonrho wanted Land Rover and Volvo had made a pitch for the bus division. Land Rover became the focus of the political storm, with 90 per cent of people polled by MORI saying that it should stay British. Despite the venerability of its main product, its production facilities had been streamlined: operations in fourteen plants had been concentrated into one, at Solihull, and that had been equipped with driverless trucks and all the latest manufacturing techniques.

On March 21, just as the talks with General Motors seemed close to completion, they collapsed in acrimony. Paul Channon had called GM's managers in and told them Land Rover was no longer part of the deal: a sale was not politically feasible. The managers retreated to Detroit in considerable dudgeon. The Land Rover negotiations with other companies teetered on for another month, before the government announced the company would stay within British Leyland. The next week, on May 1, Graham Day took over and started to pick up the bits. One of his first moves was to pretend that the company was not British Leyland any more – it was renamed Rover Group.

So the Westland affair did have a crucial industrial impact. Had it not blown up, the GM deal (and possibly the Ford one as well) would almost certainly have been pushed through – taking the British motor industry off on a very different course. As it was, in the week following the breakdown of talks, Land Rover Leyland signed a deal under which the Dutch company Daf would distribute some of its products on the Continent. It was a fruitful move.

On March 12, Elders IXL sold its entire 6 per cent holding in Allied-Lyons, leaving Elliott with a £40 million profit.

It was in the middle of the Imperial/Hanson/UB/Argyll/Distillers affair that doubts were first raised about fair play. The Stock Exchange asked Morgan Grenfell to explain why it had spent far more than it was allowed on buying shares on behalf of its clients, UB and Guinness. The merchant bank said that it had been indemnified by them – but no one doubted that it was breaking the spirit if not the letter of the takeover rules. That the whole game was starting to get dirty was highlighted by the increasingly mucky knocking advertisements in the papers. Imperial's were criticised most strongly: one showed a graph of Hanson's share price that stopped just before it would have turned up; another took a quote out of context from the *Daily Telegraph*, giving the opposite impression to the full article.

One of the dirty tricks came out into the open after a story appeared pointing out that James Gulliver had not, as he claimed in *Who's Who*, been to Harvard. It emerged that the leak had emanated from Peter Binns, the head of Distillers' public relations consultant Binns Cornwall. PR men were playing an ever more important role in takeovers. So too, unseen, were private detectives, who ferreted around in the private lives of all the important players. It was all getting rather grubby.

Hanson won Imperial on April 11: United Biscuits announced that it had only 34 per cent of the company, and was not continuing with the bid. The deciding factor was not the price. Both sides had seen their shares swept up as the *FT* All Share Index rose by 20 per cent in the first three months of the year (largely because of takeover mania), and by the time the decision was made, the offers were level pegging at £2.8 billion. Nor was the propaganda campaign crucial – small shareholders may have been flattered by telephone calls from Imperial directors but their voice was not, in the end, decisive. Nor was it even a matter of personalities. Both Sir Hector Laing, quiet and likeable, and Lord Hanson, suave and charming, were admired in the City.

Ultimately, the fund managers plumped for Hanson because he was the safer bet. He was offering rapid and sure returns by breaking Imperial up, while Laing was taking a worryingly long-term view, talking of international food groups that could take on Nestlé, Nabisco and the like. The decision came a week after Sir John Harvey-Jones had said in his television Dimbleby Lecture that "the UK has an absolute and inescapable need for international companies". Within three months, Imperial's hotel and restaurant division had been delivered to Trusthouse Forte, and buyers were being lined up for other divisions. Only tobacco, with its healthy stream of cash, would stay within the Hanson empire.

Exactly a week after the resolution of the Imps bid, Guinness won Distillers. "Ernest Saunders and Roger Seelig (of merchant bank Morgan Grenfell) have done a superb job in winning control of Distillers," said the *Daily Telegraph*. "Mr Saunders has assured himself a place in the honours list and Mr Seelig has further restored the fortunes of Morgan Grenfell."

If ever words were made to be eaten, they were those. Why Saunders bid for Distillers is a matter of dispute. His son later claimed that he was cajoled into it by Roux and Guinness' aggressive advisers at Morgan Grenfell: having tasted more than their share of blood, they were slavering for more and larger victims. But Saunders was not an easy person to bully, and there were perfectly good commercial reasons why he should want Distillers. Just as Sir Hector Laing dreamt of being a giant in foods, Guinness would become one of the giants in drink. It was a good business move, and Saunders was a good businessman.

The first suspicions about his probity came soon after he had won Distillers: he began to break the promises he had made to the Scots when the deal was first forged. No mention was made of moving the headquarters, and he showed no sign of honouring the commitment to install Sir Thomas Risk as chairman. The story of this differs wildly according to the teller. Saunders (through his son) says that Risk was obsessed with legal "details" – such as moving the headquarters – and wanted a role far more active than would be normal of a non-executive director. He was, he implies, being harried by a reactionary Scottish mafia. Risk, by contrast, says that Saunders at first avoided him then, having been tracked down in Washington, tried to steamroller him. In early July, Saunders announced that he would himself become chairman – which was a clear breach of a formal commitment to shareholders. Some Saunders loyalists in Fleet Street backed him; the institutions that had provided his money were baffled, and somewhat scandalised.

In the summer of 1986, eighteen months after flotation, Brian Taylor finally made a move: Wardle Storeys bought the conglomerate RFD for £28 million. It was the same company that Taylor had originally wanted to buy in 1979, before being sidetracked by Graham Ferguson Lacey into the morass of Bernard Wardle. "We had been looking at a vast range of companies," says Taylor. "RFD kept on coming up and I kept chucking it out on the grounds that it would be indulging a sentimental attachment to the past." What he did with it could hardly have been less sentimental: out went the textiles and cables companies, away went some property, and in came £26 million in cash. He was left with a parachute and rubber dinghy manufacturer, with profits of £2 million and sales of £20 million; it had cost him £2 million. It was as nifty a break-up as could be imagined and "next Hanson" mutterings spread through the City. Taylor was unimpressed. "I happen to believe it is more difficult to be a good manager and run things well than it is to do a break-up analysis of a company and rush around like a glorified barrow boy," he said.

Over at CH Industrials, Tim Hearley was doing some clever things too. Having been fought off by a sunroof manufacturer, Banro, in the autumn of 1985, he hung on to his 27 per cent stake, and did not sell it until the next March. The bull market had worked wonders on the shares and gave him a £1 million profit: that just paid for another sunroof company, Valor Bruce, which he bought in April. In a bull market, some things in life really are free.

Hearley was getting into his stride. In May, he paid £4.5 million for an office furniture company, Parnell & Sons. CHI's sales lifted above £30 million and profits above £2 million: Hearley, like Taylor, was becoming a City darling. But also like Taylor, he was much more than a wheeler-dealer. Ford was now fitting sunroofs as standard: Tudor Webasto boomed. Aston Martin Tickford was given a £2.25 million centre to get on with its advanced design work, and won a £7 million contract to design and supply Metro Cammell's railbuses. CHI was digging out some satisfactorily esoteric niches.

In August, the government announced that it would uphold the Monopoly Commission's recommendation that GEC should not be allowed to buy Plessey. Lord Weinstock had used all his lobbying might: members of the MMC were given tours of his facilities, and his new chairman, ex-Cabinet minister James Prior, urged and cajoled where he could. Perhaps Weinstock tried too hard, for his great opponent was Peter Levene, head of procurement at the Ministry of Defence who argued that it was precisely this sort of influence that should be limited. The problem was that in defence (where British firms still had an advantage) there were monopoly problems; in the other areas, a bigger group was almost a necessity. Never was a great British compromise so badly needed – but the Monopolies and Mergers Commission was not in the business of producing compromises.

Just to rub salt into Weinstock's wound, the government announced in December that the Nimrod Airborne Early Warning project – centred round GEC technology – would be scrapped in favour of Boeing's AWACS. More irritating yet, one of the deciding factors was that Plessey, along with Ferranti and Racal, had joined Boeing's consortium, offering to undertake much of the 8,000 man years of work that had been promised to Britain as part of the "offset" deal. The AEW project had been rather a waste of nine years and £900 million of public money.

At the beginning of September, the Monopolies Commission gave its verdict on John Elliott's proposed takeover of Allied-Lyons. The financing arrangements, it said, were not against the public interest.

## Big Bids and Politics

The import ban on leveraged bids had been lifted. Elliott had always threatened to renew the bid, even though he had sold his shareholding in March, but the bull market had now pushed Allied's price up to about £3 billion. On September 18, he rang Allied Lyons to announce that he was buying Courage from Lord Hanson instead. "They seemed rather pleased," he said. Courage would cost him £1.4 billion; that now sounded like small change.

On October 27, 1986, the City had its Big Bang. By breaking down the old-fashioned barriers between the various functions of the stock and bond markets, it was supposed to create an environment competitive enough to maintain London's financial eminence. It should, in a planned way, have done to finance what the recession had done to industry.

But it had a funny way of going about it. There were those who looked at the lavish new buildings and inflated salaries, and wondered whether, if the City had been a quoted company, it would have been allowed to get away with such extravagance. The high salaries were unhealthy in another way. If City people started to believe what their wage packets were telling them – that they were separate and superior to the toilers in the outside world – they would find it even more difficult to fulfil their function of supporting that outside world.

On November 24, a fat file arrived at the Department of Trade and Industry. It had come from the Securities and Exchange Commission in Washington, and it contained admissions from the arbitrageur Ivan Boesky, now under arrest for insider trading. They showed that the Guinness share price, the key to winning Distillers, had been artificially manipulated upwards. Boesky was one of a number of rich and powerful men who had bought Guinness shares simply to force the price up – they had been indemnified against any loss by Guinness itself. The rules of the takeover game had been stretched too far and that, it was later alleged, was why Saunders was so keen to stop Risk becoming too closely involved in the company.

A week later, on December 1, officials from the DTI arrived at Guinness' headquarters, and Ernest Saunders fell abruptly from grace. It was exactly a year since Argyll had made its first offer for Distillers.

A postscript to the Year of the City came from the man who started the whole hostile takeover boom, Sir Owen Green. BTR's experience with Dunlop had been happy: in 1985, £74 million of its £115 million of extra profit came from the company. Long-serving Dunlop executives fell over themselves to say how grateful they were for the new style. Dunlop Automotive, the car wheel division, kept the same management but increased its productivity by 50 per cent in the eighteen months following the takeover. The drawing office was replaced by a CAD centre, and a crucial contract was won – against German competition – to provide Ford with Sierra wheels. In October of 1986, BTR announced it was investing £6 million in the division, giving it the most advanced facilities in the world.

The union leaders were not so sure about the new régime. "We hit 'em hard," says Green. And, as Dunlop's managers admitted, Edwardes had already got the company to the top of the slope: Green just had to push it down the other side. He would be lucky to find another company that was quite so easy to rescue.

Green describes his acquisition strategy thus: "Every three or four years, we emerge from our place, make a large acquisition, then go back to digest it, and to grow organically." On November 20, 1986, he was ahead of schedule – it was barely twenty months since he had bought Dunlop – but decided that in any case it was time to move. Perhaps he wanted to have one more triumph before handing over the reins to his chosen heir, John Cahill. BTR bid £1.1 billion for Pilkington, the glass company.

Pilkington, established in 1826 and still headed by a member of the family, was one of the great "one town" companies in Britain. It was St Helens in Lancashire, and St Helens was Pilkington. Unusually, though, it managed to combine paternalism with the most modern strategic thinking. Its recent fortune had been based on float glass technology, which it had invented in 1959 and licensed around the world. But by 1980, when Antony Pilkington took over, its earnings had flattened out and it needed somewhere else to go. Pilkington, fifth generation of the family and former lieutenant in the Coldstream Guards, was a thoroughly modern businessman. As the glass market slipped, he had cut the workforce, pushed management control downwards, and uprated the marketing effort. He moved into continental Europe, buying a majority stake in West Germany's biggest flat glass maker in 1981, and then taking over one of America's major safety glass companies. With the biggest share, 18 per cent, of the world flat glass market, Pilkington was one of the few British companies with real strength; it had, the *Financial Times* commented, "done everything a management consultant or business school professor could demand".

As always, Green was being opportunistic. The glass company had yet to gather in the rewards of its virtue, and had been burdened for year after year with heavy redundancy costs. Nevertheless, it marketed that virtue brilliantly in the battle. David Challen, Williams Holdings' mentor from the merchant bank Schroders, orchestrated the campaign: it was long-termism against short-termism; the north against the south; industry against the City. Green was taking on more than a company – he was assaulting a community and, with the Guinness scandal frothing furiously in the background, he could hardly have chosen a less politic time to do so. He was a thoroughly black knight and Tory MPs, their eyes on an imminent general election, hoped he would quietly go away.

That was what he did – although it was unlikely that politics influenced him much. In mid-January, Pilkington produced a profit forecast of £250 million, higher than anything the City expected and twice the previous year's figure. As its shares rose way out of sight of BTR's bid, Green had to decide whether to raise his offer: the price was now, it was clear, close to £2 billion. On January 20, he said he would not. There were parties in St Helens.

The defeat marked the end of the great British hostile takeover run that had been started by Green himself with his bid for Thomas Tilling. There was a purely financial reason why opposed bids now dried up. With City "sentiment" against them, the predators found that their price/earnings ratios were often below those of their targets. But there was a commercial reason too: Britain was running out of really badly managed companies, at least among those quoted on the stock market. In acknowledgement of that, the predators started to look elsewhere for easy pickings – mostly, they went to America. BTR, determined to be different, went hunting in John Elliott's homeland. BTR Nylex, the biggest industrial company in Australia, would make the running from now on.

Even as the British looked abroad, so foreigners were looking at Britain; in particular, the trickle of Japanese investors was turning into a steady flow. Managers watched with interest to see how the Japanese themselves would turn their theory into practice. How, especially, would they cope with the bolshy British worker?

# 18

# *People power*

Every morning at five to eight, the receptionist at Komatsu UK pressed a button, and an upbeat synthesised rhythm surged through the factory. A woman's voice started giving commands: "Arms up, two, three, arms down, two, three..." The 350 people in the factory, including the entire management, started jumping up and down. All were dressed in identical blue uniforms. Four minutes later, the music stopped, and the workers went off to build excavators.

When the Japanese arrived in Britain, they did not directly impose their culture. Instead, they appointed a handful of senior locals, sent them to Japan, and told them to choose the techniques they liked. Invariably they took home as much as they could. Komatsu, which set up at Birtley, County Durham, in 1986, was unusual in its use of physical jerks; no one yet had a company song. But the basic principle of Japanese manufacturing – that everything came down to using people properly – was imported unchanged.

Japanese and British industry had links that went back to Japan's opening-up to the West after the Meiji restoration of 1868. In the 1870s, a Scot, Henry Dyer, had become the first head of the Imperial College of Engineering, and a stream of British engineers taught the Japanese enough to set their country up as a great military power within fifty years. Even as the status of engineers slipped away in Britain, Dyer and his like were revered in Japan; his name is still honoured today.

But, by the early 1970s, the balance of economic power had shifted sufficiently for Japanese companies to consider Britain as little more than a useful platform for European manufacturing. The first Japanese enterprise to set up was YKK, the zip company. That was in 1971. Companies continued to trickle in through the 1970s and by the end of 1981 there were thirty, including all the big television companies. All, at this stage, were simple assembly plants. The best known was Sony, which had its factory at Bridgend in South Wales.

The Japanese had two motives for arriving. On the one hand, it was part of a natural shift towards multinationalisation, echoing American moves in the 1950s – to get closer to markets, and simply because companies were starting to outgrow Japan (the jump in the yen from 1985 gave further impetus). On the other hand, it was a defensive move. The Japanese became increasingly alarmed by the protectionist noises emitting from Washington and Brussels; they were acutely aware that their own government was in no position to give anyone lectures about freeing up markets, so they decided the best bet was to leap the wall before it was built. From 1984, the prospect of a Fortress Europe rising up from the "1992" programme haunted them further. They had already moved in on Latin and North America; the logic of establishing a presence in Europe was overwhelming.

But where to go? The Japanese, isolated for so long, felt a need to huddle together when they travelled abroad. They had settled on Germany, and particularly Düsseldorf, as their favourite base for sales operations. But Germany was too expensive to manufacture in, and language was a problem. In a factory, they would have to communicate more with the locals: they were much happier speaking English. Low-wage Britain was the obvious solution, except for one thing – Eikokubyo, the English Disease. The Japanese had had their own period of industrial turbulence after the war, which had been ended by the defeat of a great strike at Nissan in 1953. Since then, industrial relations had been dominated by company unions, which relied on cosy relationships with the bosses: Japanese managers had no experience of direct confrontation, and would do anything to avoid it.

But the industrial relations experience of the pioneers who braved the brickbats was remarkably good. The Sony plant was held up as a shining example: people said that the Welsh and Japanese got on particularly well. In fact, the new order of Japanese companies, in which confrontation was replaced by paternalism, would probably have been gratefully accepted by conflict-weary British workers anywhere. One notable failure, the Toshiba/Rank television joint venture at Plymouth, was attributed to the fact that Toshiba allowed the British to run the plant as they wanted.

The arrival of the Tories in 1979 triggered a new phase of Japanese interest. Not only were the unions being satisfactorily suppressed, the government was clearly determined to offer whatever blandishments it could to lure companies in. This determination was based not so much on a wish to create employment as on a belief that a few big Japanese factories would give idle British managers a stirring example: bring in the school swot and buck the other chaps up. The Department of Industry's Invest in Britain bureau went on a fishing trip in the East with some serious tackle; it was backed up by the British embassy in Tokyo, where a potent collection of commercially astute diplomats had been concentrated.

In 1981, British civil servants started dangling carrots in front of the second largest Japanese car company, Nissan. London would be prepared to offer almost any assistance if it agreed to set up a factory. Representatives from local councils throughout Britain flew to Tokyo to pay homage to the master mechanics, and to describe the joys and grants of their piece of Albion. Most important of all, the British were able to describe a new type of industrial agreement which, they said, could resolve all worries about Eikokubyo.

Roy Sanderson was the national engineering officer of EETPU, the Electrical, Electronic, Telecommunication and Plumbing Union. In the 1960s, he had been a communist shop steward at Lucas. Now, he was a convert to moderation, and had been particularly impressed by his several visits to Japan. He had also studied the effects of British strikes in 1979 and 1980, and had concluded that the strikers rarely ended up better off. It was rational, he decided, to trade the right to strike for other changes, principally single status working and some sort of industrial democracy: an advisory but influential union/management council was his favoured solution. The workforce should be more flexible, he thought, but companies must be prepared to offer training to achieve that flexibility. And as a backstop, a system of "pendulum arbitration" should be used to settle disputes: the arbitrator had to choose one proposal or the other – he could not suggest a compromise. That, inevitably, forced both sides to moderate their demands before presenting them. Sanderson had invented the no-strike package.

The Japanese had nothing quite like it, but it provided exactly what they wanted in a union. In 1981, the EETPU signed its first no-strike agreement with Toshiba at the site of the failed joint venture plant in Plymouth. Toshiba had bought Rank out, and was ready to apply a Japanese solution.

The British blandishments stirred up a byzantine power struggle at the top of Nissan. Officials watched amazed as photographs of the head of the company union living it up on his yacht appeared in the Japanese press. They were, it emerged, part of a smear campaign that ended the career of Ichiro Shioji, a man who had assisted at the creation of the new style of Japanese union. He had helped break the 1953 strike, and had at times been the most powerful man in the company. He was opposed to the British plant; the chairman and victor in the battle, Takashi Ishihara, was in favour of it, and in January 1985 he announced that it would indeed be built. For those who remembered when Nissan produced Austins under licence in the early 1950s, it was a stirring moment.

The site chosen was in Washington, near Sunderland, an area devastated by the recession and with a huge pool of unemployed labour as well as maximum grants. In April 1986, the first European-built Japanese cars rolled off the line.

Nissan was just the best known of the 1980s Japanese implants. It was also the most politically sensitive, and stirred other European governments into paroxysms of rage – the French at first refused to accept the Bluebirds as British, and included them in their tiny quota of Japanese imports. European motor manufacturers were happy to admit that they were not yet ready to take the Japanese on in a fully open market, and saw companies such as Nissan as Trojan horses inside the European Community. Many continentals, particularly the French, had always treated the free market as something of a luxury, to be quietly shelved if the going got tough. Now, they demanded that it be put firmly at the back of a cupboard. Even the European Commission, which was so firmly committed to free trade in principle, found it remarkably easy to prove that Japanese firms were dumping their products in Europe at below their proper price. The British, of course, could afford to take a high moral tone in all this – they were the chief beneficiaries of Japanese investment.

Under political pressure, the British had to insist that Nissan buy an escalating percentage of its parts within the Common Market. European manufacturers were determined to prove that all Japanese factories were "screwdriver plants", assembly operations that would import as much as they possibly could from Japan, and thus provide little benefit to local economies. Until the mid-1980s, that had been largely true, although in 1982 Sony, which had been struggling to educate its suppliers to provide the quality it needed, had set up a factory to make picture tubes, thus increasing its local content to 90 per cent.

The new wave of arrivals, led by Nissan and Komatsu, all realised it was expedient to be seen to be buying as many parts in Europe as possible. Komatsu's Japanese managers complained bitterly about the quality of British supplies to start with but, by applying their own techniques, they slowly built up an acceptable base of subcontractors. Their techniques were broadly those used by Ford with Tallent, but could sometimes be taken to extremes. Before choosing a subcontractor, the Japanese managers would want to know every detail of his manufacturing technique, how long it took to do each process, what his raw materials costs were. Then, having awarded the contract, they would visit their supplier many times a week, or even day, making suggestions, cajoling, checking up. "You could call it support," said a disillusioned ex-employee of one of these firms, "I would call it harassment." But it worked for the Japanese: they got their quality.

The area that benefited most from Japanese investment was the North-East of England, the Newcastle/Sunderland/Durham area, which drew in twenty-five Japanese companies in the second half of the 1980s. Its initial attractions were grants and an ample supply of labour; later, the Japanese herd instinct took over.

Seminars, gurus and management consultants could explain in detail how Japanese manufacturing systems worked, and the theory was splendid. But what they could not explain was how the Japanese themselves operated, and where fact and theory did not match. That was where the lesson of the likes of Komatsu and Nissan were so useful. How did the school swots measure up in the classroom?

On the whole, pretty well. At Komatsu, incoming calls were always answered immediately. Everyone, including the senior managers, wore identical uniforms. No manager had his own office; there was even talk about rotating jobs between office and shop floor. Quality circles were strong, with a fairly sophisticated level of statistical analysis expected of their members. An advisory council, set up on Sanderson's lines as part of a single union agreement, successfully avoided disputes. On the shop floor, great yellow machines were being assembled; a sign read "Always remember the next station is your customer" – the total quality theory. One of the assemblers was sweeping up around his machine. There was just one piece of advanced equipment – a £1 million Computer Numerically Controlled machining centre. It was all as the textbooks said it should be.

The majority of workers were content. Thanks to a high level of overtime, they earned more than they could in local British-owned factories. The efficient receptionist claimed that this was the best job she had had. There was a gym, should the morning work-out not prove rigorous enough.

In Japan, the big companies built up a core workforce, the size of which was geared to the bottom of the demand cycle. As demand rose, temporary workers would be drafted in; as it fell, they would be laid off. That was how guaranteed lifetime employment worked: those lucky enough to be selected for the core became part of an élite, spending their lives in company housing, going on company holidays and being pampered when they retired. Those who were not chosen had a tougher life. But there was a price even the core workers had to pay: they would initially be selected as much for their orthodoxy as for their ability, and if they strayed from that orthodoxy (for instance, by marrying an unorthodox wife), they could find their lifetime employment terminated.

This was all a bit big brotherish for Britain, but there were elements that could be extracted. The idea of having a core workforce was one that managers throughout the West were looking at seriously: companies had tended to become overmanned mainly because it had been difficult to reduce staff when the market turned down. In America, the idea of "temping" was being stretched beyond secretaries to manual workers, even to engineers and accountants, while in post-recession Britain companies had taken to hiring self-employed workers on short-term contracts whenever they could. From 1983 to 1985, temporary employment grew by 11 per cent.

The Komatsu solution was to create a core workforce, and use overtime to cope with strong demand. That had two advantages: it kept the core as small as possible, and it kept its workers' pay high. As the company expanded strongly in the late 1980s, they often worked twenty hours overtime a week. But Komatsu (and all the other companies that arrived) had an advantage that would eventually disappear: a vast pool of workers in which to fish. When it advertised its first seventy production jobs, it received 3,000 applications, and another 4,000 had applied before full strength was reached. It could afford to be just as choosy as the Japanese companies back home. Two and a half hour psychometric tests were given to anyone wanting a supervisory role, and more elementary tests to others. "Personality and attitude" were more important than qualifications and experience. The vast majority of workers were in their early 20s: easy to mould, cynics said. Security at last, said those who passed the tests.

Nissan was bringing a whole new type of factory to Britain – a full-scale mass production line. Anyone who had read a book called *In the Passing Lane*, which appeared in English in 1983[1] might not have found that comforting. In it, a Japanese journalist, Satashi Komata, described his experience as a temporary worker on the Toyota production line: it consisted of little but relentless and near panic-stricken pressure. But the description was no worse than that of the Ford line at Halewood in the 1960s, and the actual experience of people who worked at Washington was, it seemed, a lot better.

The managers at Nissan – predominantly British – made a point of saying how different their plant was from those in Japan; it was certainly less Japanese than Komatsu's. For example, target daily production figures were used, and an indicator showed whether the line was hitting them; that was more American than Japanese. And while the workforce was divided into twenty-strong teams (like cells), the supervisor of each was responsible for hiring new members of that team. Peter Wickens, Nissan's director of personnel, had already used this idea at his previous employer, the American company Continental Can.

Nevertheless, the basic philosophy of pushing responsibility downwards came straight out of the East. Each day started at 8am with a five-minute team briefing, when the supervisor discussed problems. (This was backed up with weekly information bulletins and twice-yearly mass meetings.) Then the line started. As the cars edged from station to station, two men – probably between 20 and 25 – carried out a series of tasks; they had four minutes and forty-eight seconds to complete them before the next car arrived. Many of the tasks were strenuous, and if anyone had a problem, other team members rushed to help him out. If a worker made a mistake, it was his duty to mark it with a label, so that it could be corrected. He would not be told off – just, if necessary, given more training by the supervisor.

There was one check and repair man in every team, but quality control was fundamentally in the hands of the assembler, with the next man down the line, his "customer", there as a back-up. Faults had to be cured before the car left the team area: end inspection was very limited. The workers were encouraged to think of better ways to do their jobs; as long as the supervisor approved, they could use any technique they liked. The team system worked well and, despite the physical toughness of the work, no one seemed to have any complaints. As the Washington plant had been acknowledged as building some of the best quality Nissans in the world, nor did the company. It was not a bad achievement for the sloppy British.

One of the most successful features at Nissan was the emphasis given to the role of the supervisor. In Britain, the foreman or first line manager usually had an unenviable position. He would probably have been promoted from the shop floor – not because he was likely to be a good manager, but because he was a good shop floor worker. He would not be trained properly and would find himself isolated from his erstwhile workmates, while still not being "true" management. Increasingly, his role as a manager was usurped, by specialist departments such as personnel on the one side, and by shop stewards on the other. He was like an NCO without power or prestige.

Some companies, including Ford and Jaguar, were trying to expand the supervisor's role, and give him better training: they even tried, with little success, to attract graduates to the job. For Nissan, the supervisor was pivotal; he acted as a mini managing director for each team, ultimately responsible for its quality, costs and production. The company went to extraordinary lengths to pick the right people, who were then paid much more than they would have been in an "ordinary" company. The selection process (to pick twenty-two people from 3,500 applicants) included extensive psychometric tests, group exercises and even an informal dinner: it was an extraordinary way to pick a bunch of foremen – but it worked.

But there could be problems with the Japanese way: the chief one was simply that Japanese and British cultures did not mix very well. One British manager was offered a senior job on a good salary when Komatsu first set up in 1986. He moved his family up from the south of England but, when he reported for work, he was greeted with blank non-recognition by two Japanese managers: they said that they had been expecting a clerk, and each claimed to have his job. It would have been farcical had it not been so frustrating for the Englishman, who could only watch as the other two battled it out in Japanese. Eventually, the one with better contacts in Tokyo won, and the loser, along with the British manager, was transferred to another department. Finally, the Englishman found himself assigned to sweeping out the back yard; he took the hint and left.

He concluded that he had become trapped in a battle that he could not possibly win, because the system of patronage meant that if there was a dispute, the man with a direct line to his senior would win. That could happen in any company, but here the patronage system was strictly inter-Japanese; the failure to demarcate jobs clearly was also in the Japanese tradition. Even more baffling was the Japanese system of consensus, by which decisions would emerge by an osmotic process unfathomable to the outsider. At meetings, for example, there would never be an agenda, no minutes would be taken and non-Japanese managers would often leave wondering what, if anything, had been decided. This was not a big problem for most, who were left to get on with their jobs, and the biggest firms never had more than a handful of Japanese on site (Komatsu had eleven permanently). Inevitably, though, as companies continued to arrive from the East, more locals would find themselves tumbling down the cultural chasm.

For workers on the shop floor, Komatsu was about the nearest thing to nirvana to be found in a British factory. As the managing director, Torio Komiya, wandered around picking up pieces of paper, they would chat to him. When he made a speech at the annual dinner/dance, he was greeted by cheers and catcalls: the barrier between shop floor and management seemed genuinely absent. In a fully working Japanese system, that was as it must be. The whole basis of true Japanese systems – total quality, quality circles, even just-in-time – was the transfer of responsibility, and therefore power, from the management to the workforce. In a full just-in-time factory, there would be no cushion to allow for the failure of an operator to do his job properly: if he stopped work, the whole line would grind to a halt almost immediately. At Komatsu which, by its nature, worked at a slower pace, this urgent requirement was missing but, as each operator had to "sign off" his work, the principle of individual responsibility was just as strong.

There was one fundamental reason why Andy Barr's strategy at Austin Rover or the MRP system at Leyland Trucks was "pre-Japanese". Neither relied on the total cooperation of the workforce. In the early 1980s, with management at British Leyland at its most macho, that was probably just as well. But, as the need for that machismo faded and a *pax disoccupationis* – unemployment peace – ruled, the more thoughtful managers looked around and wondered if there was some way of bridging that sinister chasm between manager and managed. It was a chasm that had widened threateningly during the recession.

The statistics showed that strikes had virtually disappeared. By 1987, the number of working days lost through strikes in manufacturing had tumbled to fewer than 600,000 – compared with 22.5 million in 1979. It was the most peaceful year for half a century. The onset of industrial peace had come suddenly. In 1979, 29.5 million days were lost in all; in 1980, 12 million; in 1981, 4.25 million. From then on, such strikes as there were were concentrated in the public sector: civil servants (1981), health workers (1982), water workers (1983), miners (1984 and 1985), and teachers (1985 and 1986). In manufacturing only the occasional flare-up in the motor industry marked almost perfect peace.

But was there harmony? The obvious conclusion was that the threat of unemployment was behind this improvement, backed up by the government's laws. Many managers found the atmosphere after the recession stimulating, but for workers who had no control over their fate, it was a time of uncertainty and depression. There were plenty of factories, even quite big ones, that had been the comforting centre of communities, where generation after generation of the same family had worked. The lorry plant at Leyland was one. Whatever was now being said about quality, the workers felt well into the 1970s that they were producing the finest trucks in the world. "We had pride in the company and in the quality of the products," says shop steward Ian Hays. It was only when the cuts started, in 1979, that that disappeared. "Morale went underground," says his colleague Barry Morris. "Every time the company made an announcement, we expected it to be about more redundancies."

With hardly an engineering factory unaffected by job losses, relations between what was left of management and workforce had become strained. "From 1979, the trade unions were put in prison, and the government was expecting more worthwhile citizens to be released," says Hays. "It doesn't happen like that. Attitudes harden."

On March 29, 1982, the London Chamber of Commerce held a conference entitled "American union avoidance techniques". Its speakers were "union busters" who offered a package of legal advice and training to managers who wanted to force unions out of companies. In 1981, 848 "decertification" elections were held by the United States Labour Relations Board: in three quarters of them, employees voted to get rid of unions. There was, the Americans said, very strong interest from British companies.

Apart from the odd cause célèbre such as News International, only a handful pursued derecognition as a policy. But there were sufficient complaints from unionists about "macho management" to make it quite clear that many companies were taking full advantage of the weakness of the unions to push their workforces about. They were encouraged by the government, which introduced increasingly tight restrictions on the right to strike. It could point out, accurately, that it was only bringing British law into line with common practice in Europe. Even after the three major pieces of anti-union legislation, unofficial strikes were not illegal, as they were in Germany, for instance. But that was an academic argument: the laws did little to rebuild trust between workers and management.

The Japanese did not have to rebuild trust, just build it: that was their big advantage. As outsiders, they could sidestep the "us and them" problem in a way that no Briton ever could. British managers, meanwhile, had a choice: either they could do nothing, and hope that the balance of power would stay in their favour indefinitely; or, as the chopping stopped, they could find some way of bringing the aims of the workforce and the company into line. It was an obvious aspiration but a difficult one to achieve. Many different approaches were taken and, by the end of the decade, it was far from clear which, if any, were succeeding.

Attempts to analyse exactly what was happening in industrial relations were further complicated by the great shifts taking place in the trade union movement. Between 1979 and 1987 union membership fell from 12.2 million to 8.8 million. The Transport and General Workers Union lost a quarter of its members, half a million people. Some unions, unable to cut their overheads in line with tumbling subscriptions income, faced bankruptcy. Mainly, membership fell because so many jobs were lost in traditional unionised industries, while new jobs were being created in service and high technology industries where unionisation was much weaker.

But there were other reasons. Companies were able to insist that bargaining should be carried out at local or company level – the old national pay agreements with employers' associations, where union muscle was most effective, started to disappear. More fundamentally, the sweep towards individualism – tapped but not created by the Thatcherites – was making the old collectivist union ideas seem out of date.

The watershed for the unions was the 1983 election. Despite the level of unemployment, and ignoring explicit advice from their leaders, fewer than 40 per cent of unionists voted Labour, 12 percentage points fewer than in 1979. Just as some managements responded to the pressures of recession and others did not, so some unions tried to ignore this vote of non-confidence, while others adapted. As with companies, there was a spate of defensive mergers but more fundamentally, union leaders started to look for a new raison d'être. When they found it, they called it "new realism". Instead of waging class war, the unions would go back to their nineteenth-century roots, when they were principally welfare organisations. Many still provided death and sickness benefits, and it was these that would be expanded to take in modern-day versions such as life insurance and discount holidays. A new rôle, training, was added. Two new-realist unions, the Amalgamated Engineering Union and the EETPU, both set up technical training colleges.

The personification of new realism was Eric Hammond of the EETPU. He stood for everything that Arthur Scargill hated, rejecting the old class-based antagonism in favour of cooperation with management. Not surprisingly, this approach was seen by old-style unionists as a complete negation of collectivism, of what unions were supposed to be about. The EETPU allowed its members to be mailed, and to be offered discounts on everything from financial services to car breakdowns; it was becoming more like the AA than a traditional union.

Behaving exactly as though it were a reviving industrial company, the union marketed itself vigorously, both to potential members and to 500 employers which, in 1984, received a fourteen-page glossy prospectus: it showed pictures of its high tech training centre, complete with robot. That year, too, the EETPU held a fringe meeting at the CBI conference and Hammond, tongue in cheek, suggested he be allowed to join the employers' organisation. He loved riling the old-style unionists, achieving pariah status both by refusing to back the miners in their strike and by taking government money for postal ballots. They finally lost patience with him in 1988, and threw his union out of the TUC. By then, though, most of them were following along his path.

During the bad days of the recession, it suited companies to put on an enlightened mask and talk loudly about "communication". Michael Edwardes' "Blue Newspaper" was an example of this new communication – it meant bypassing the unions to explain to the workforce just how bad things were. That was fine while they were bad but, as profits started to become embarrassingly healthy, some companies quietly dropped these efforts: they would, they thought, only encourage big pay claims. In any case there were now more important things to think about than industrial relations.

Inevitably, it was those companies that were following the Japanese path that took the opposite view: that now was just the time to try to establish a reasonable basis for the days when unemployment fell. In pursuit of Japanese lore, they took communication ever more seriously: they saw it as a tool to break down the barriers, giving it a wider meaning that included an element shamefully ignored by British business – training.

In 1986, Austin Rover decided it was time to soften its macho image, and move further towards the Japanese way. Cell working had already been introduced, and workers – armed with statistical process control – were "signing off" their own work after self inspection. Now zone circles, similar to quality circles, were introduced along with briefings which involved stopping the track once a month while the foreman would give a talk on the performance of the zone and the company.

Other devices to aid communication – closed circuit monitors on which both information could be given, and workers could put advertisements – were all supposed to help the team spirit. The company also tried to introduce a sense of belonging to new recruits: they would be asked to come in at the weekend with their families to look around. The recruitment took on a Japanese flavour: candidates would be asked to come in for two days of discussions and tests. As well as discovering whether they had the right skills, the aim was to find out whether they were likely to identify with the "philosophy" of the company.

Up at Leyland, John Gilchrist also started to move the lorry company in an easterly direction, introducing a "quality improvement process" – a total quality programme – in 1985. The basis of this was a training programme which explained to everybody, not just on the shop floor, what the principles of total quality were, and how they fitted in. Quality circles were set up, SPC installed. Then the workers were allowed to get on with it: they had to sign a build sheet, saying that they had done each job properly. The quality of the vehicles was measured on a "zero defects" basis, that is, any fault, be it only a speck of dust in the paintwork, counted as much as a missing wheel: subjective judgment of what was a fault was thus banished. Final inspection was limited to an audit team, not on the factory's payroll, taking an average of one and a half vehicles a day (out of seventy-three produced), and going through them with a fine tooth comb. If a fault was found, the worker who had done the job would be asked what had gone wrong. More often than not, the problem would be traced to a component, and the supplier would be castigated.

In Japan, suggestion schemes had become the rage. Leyland borrowed the idea, pinning up photographs and having presentation ceremonies for the staff who made the best ones. Gilchrist was surprised how well it appeared to be received. Ian Hays was less impressed. "We don't live on patronising gestures," he said. "We will make a profit and when it comes to sitting with the gaffer, we will want money, not a mug. We don't like the Come On Down attitude."

Nevertheless, quality did improve, machines stopped breaking down, and parts started to arrive on time. Gilchrist believed that these changes were psychologically important. "Morale," he said, "comes out of an orderly workplace." In the case of Leyland, it also came out a merger with Daf. Barry Morris went on a course at its factory at Eindhoven in Holland, and came back nostalgic. "Their assembly line was like ours fifteen years ago," he says. "There was a degree of trust between the workforce and management." British industry still had a long way to go.

One final example of Japaneseness is interesting. It is the only example of a Japanese company taking over a going British concern, in a dreadful state and complete with fearsome unions. It continues the story of Dunlop.

Sumitomo took charge of the sprawling tyre factory at Fort Dunlop on January 1, 1985. Within three years, output had risen by 50 per cent, and productivity had doubled. This is a story that must be treated with caution, because Sumitomo took over at the best possible time. There had been wholesale slashing of tyre capacity, and the market was just starting to turn up. Furthermore, the dirty work had already been done: the workforce was down to 2,000 and morale had only one way to go – up. Nevertheless, it seems probable that Sumitomo made more of the factory than Dunlop would have done.

The Japanese treated the company with kid gloves. Day-to-day running was handed over to a British managing director, Gerry Radford, helped by Ian Sloss, who had been at the Fort since 1979 and was now in charge of production and personnel. There were few physical signs of Japanese ownership. Grey company jackets were optional: some managers would put them on over their suits – an incongruous mix. Fort Dunlop still looked shabby, it still had its toytown fire and ambulance station, and plenty of workshops full of nothing much. An invisible line zigzagged through the vast site, marking SP Tyres (as the company was now called) off from the non-tyre territory of BTR. Inside the factory, lots of blue and yellow paint, and strips of rubber flowing slowly along between massive machines, before being pressed into great truck tyres. There were ten modern machines – each worth £300,000 and brought in specially from Japan – but no sign of high tech anything. The layout, the managers pointed out, was far from perfect, and investment was modest, at £10 million a year. The baptism of new machines with saké did not, apparently, make them work better.

But there were changes, plenty of them. In the best Japanese style, they related almost entirely to people. When Sumitomo took over, all union agreements were retained, making it without doubt the only Japanese company in Britain that tolerated seven unions. But, Sloss said, the number of unions itself was not important, what did matter was how many separate negotiations had to be undergone. After chiselling away, Sloss got the number down to two – and was aiming for one harmonised agreement. The payment system was revised so that 10 per cent of all profits were shared out between the workers. A communications system was set up based on monthly meetings, with information cascading down through the workforce. Managers started to wander around the shop floor, as did Japanese technicians pushing constant tiny improvements wherever they could.

The Japanese also tried to break down the barriers between the different functions by setting up "action teams" to work out objectives and solutions in areas such as supplier relationships and marketing. These were made up of people from various disciplines. "One team head, a technical manager, was introduced to an engineer," says Sloss. "They had both been here for twenty years but had never met." It reinforced what the Japanese said to him: "You have effective units, but we are like one rock."

The biggest single change was the introduction of mass training. First SPC was introduced, and was backed up with training for 800 operators. Then a total quality programme was launched. A schedule was drawn up to train everyone in the factory. In Japanese style, all training was done by line managers. To get it started, six teams of managers were formed: each had to produce a strategy for the programme, and a training schedule was drawn up by which everyone in the factory would be trained in problem-solving techniques.

Ironically, the company that gave the first and biggest impetus to the introduction of Japanese techniques in Britain was still struggling with them long after its protégé suppliers were up and running. Ford was having great difficulty in getting the unions to accept flexibility and even as its troublesome Halewood plant started to come right, the Dagenham factory was causing problems. One old factory, which used to house Brigg's Motor Bodies, was just too small to rearrange to give the best flow. Judging by the graffiti on the obligatory "quality" posters, the workforce was hanging on to its cynicism.

Japan was not the only country that British companies could borrow from. When it came to motivating workers, America's closer culture could be a better bet. The most straightforward form of motivation was financial, and the most obvious way of bringing the aspirations of the company and its staff together was to link payments to profits. This could either be done through direct profit sharing or indirectly through share option schemes. Many companies had been running these since the 1970s. The BOC scheme, for example, was started in 1976, and allowed employees to save for between five and seven years before being offered shares at a price agreed at the outset. As long as the share price had risen, they could sell immediately and make a profit. As with the share options granted to senior executives (which became steadily more generous through the 1980s), the direct link with profits was tenuous. During a bull market, it was exaggerated; in a bear market, depressed.

More overtly American was what Leyland shop steward Ian Hays called the Come On Down school of management. It gained favour in the most unlikely places, including that revitalised bastion of tradition, ICI. By 1985, Tony Rodgers felt the Organics Division was ready to be pushed further towards marketing. He changed the name to Colours and Fine Chemicals, introduced a cheerful new logo, and launched a campaign called Fighting for the Customer. He discovered that many employees had no idea who was buying their products, and arranged to swap factory visits with a number of customers, as well as carrying features about them in the house newspaper.

The core of the campaign was a competition called Hollywood Stars. Even though Rodgers had spent four years in Canada, he was worried that this was too razzmatazzy: "My first response to it was that it wasn't for us, then I changed my mind: it was sufficiently outrageous to work." Every employee was sent a letter in a glitzy *Reader's Digest*-style envelope inviting him or her to put forward marketing-based suggestions. The panel of judges included customers, and televisions and videos were handed out as prizes. The final grand prize was a trip to Hollywood. The winner was a trainee at the Grangemouth plant in Scotland, who suggested that an articulated lorry should be kitted out as a travelling exhibition centre. In due course, and £200,000 later, the Showcruiser was presented. The razzmatazz was kept up all along: Rodgers arranged for a runway at Manchester airport to be closed for ten minutes while the vehicle belted along it. Lord Magowan, ICI's grand old founder, would have turned in his grave.

Down at Havant in Hampshire, the suggestion that the 1980s brought a completely new style of management/worker relationship would be greeted with a shrug. To employees of IBM UK, the techniques being jammed into traditional factories were very familiar. Japanese techniques had, after all, originally come from the USA, and IBM was one of a handful of companies that had adopted them in those early days. Its style was a recognisable cousin of the Japanese one but, like some long isolated tribe, it had developed its own characteristics.

IBM UK had not had a strike since it was set up in 1951, and it commanded immense loyalty from its workers: in 1985, its 18,000 staff had spent an average of 11.5 years with the company. Unlike the Japanese, it would not negotiate with trade unions. This, managers said, was because they tried to emphasise the importance of the individual and to have a union would mean that they had failed as managers. They described the subtle mix of individualism and a pride in the whole as creating "the biggest boy scout troup in the world". IBM was paternalistic but, while insisting on single status in working conditions, it had a very Western belief that the most able should be able to accelerate rapidly. Pay was adjusted according to individual merit: a spirit of competition was encouraged, not suppressed. And disputes were always dealt with on an individual basis – a comprehensive set of procedures made it easy for workers to air their grievances.

This style of management had spread ever wider in the USA, where unionism was in steady retreat. But when it made it across the Atlantic, it was only to the high tech sector. With so many electronic companies founded by people who had worked at IBM, Hewlett Packard or one of the other American firms operating in this way, it was logical for them to take the system with them. One of the reasons the EETPU had to market itself so heavily (and take over jobs like printing the *Sun*) was that its "own" area, electronics, saw little need for unions.

But, as had been proved in America, there was no real reason why the IBM style could not work in other sectors. It begs the question: why, given that IBM had been operating successfully for thirty years, did British managers make no attempt to follow its example? Why did they wait until the Japanese, with an arguably less suitable system, appeared? Was it just the strength of the unions? Or was it more because they did not have to – and when they did, it was the Japanese who had the highest profile? Probably a bit of both, but it still seems strange that managers were happy to trot off half way round the world when a trip to Hampshire might have served them at least as well.

# Part Six

# *Rising and falling*

19

# Crash? What crash?

When the stock market plummeted on October 19, 1987, the City assumed that industry would crash too. In part, this was because economists believed the American economy would be plunged into recession – memories of the 1929 crash lingered. More, it was due to a feeling that if the City was having it tough, industry must be too.

Up above his car showroom in Derby, Nigel Rudd of Williams Holdings watched, amused. "People in the Midlands saw their jobs were safer than they had been for ten or fifteen years," he says. "They couldn't understand what the fuss was about." Rudd had been expecting the stock market to fall and had been cutting down debt and building up cash. Two months after the crash, Williams was off buying again: Berger was snapped up, added to Crown Paints, and the group set about creating the second largest, and most efficient, paint maker in Britain.

Industry was unperturbed simply because it had never had it so good – at least since the unreal days of the early 1970s. From 1983, the economy had been dragged along at a cracking pace by consumer demand: a cheerfully spendthrift society shelled out 30 per cent more in the shops in 1987 than it had in 1981. Not only had wages kept ahead of inflation, people were prepared to borrow more. Their houses were more valuable and, with all credit controls having disappeared in 1982, banks and other companies had been busy flattering them with the promise of loans and credit cards. In 1986, household expenditure exceeded income for the first time since records began. So it was not surprising that, in June 1987, Mrs Thatcher swept back to power for the third time.

It would have been a wild generalisation to say that every industry was blooming. In pharmaceuticals, British companies were among the strongest in the world – Zantac had transformed Glaxo into an international giant, and the £500 million research campus it was building in Stevenage was likely to reinforce its position. Chemicals were healthy, with ICI now the most profitable of the world giants. British Steel was becoming more muscular by the month. The textile industry was looking in better shape than it had for many years: Coats Viyella and Courtaulds were the two largest textile groups in Europe, and profitability and productivity were both rising fast. Electronics was slowly coming back to life after the shocks of 1985: Plessey bought Ferranti's semi-conductor business, and looked ready to take advantage of the recovering chip market.

Even the motor sector was starting to look healthier, as sales surged above 2 million a year for the first time in the UK, and brought demand within sight of capacity. Inevitably, that pulled the Midlands back on to its feet. Its economy had collapsed most spectacularly in the recession; now it was resurrecting itself with a speed that was almost as astonishing.

But elsewhere, the picture was more mixed. One part of British Aerospace was booming on the back of defence exports; but the civil aviation division was losing so much money because of the weak dollar that the whole group dipped into the red in 1987. And some sectors were still undergoing painful restructuring: even if the predators could no longer find "soft" and badly managed targets, the other sort of merger – to create larger and stronger units – continued. In mechanical engineering, for example, the bull market actually increased takeover activity in 1987, and led to a second wave of sometimes brutal shake-outs: APV bought the food machinery company Baker Perkins, Weir bought Mather & Platt, the pump manufacturer, and no fewer than six machine tool companies changed hands. It was in heavy engineering, too, that the last great piranha-swallows-whale takeover took place: the aggressive electrical company FKI (sales £84 million) bought Babcock International (sales £1.2 billion) and within a year, twenty-five factories had been closed and 4,000 jobs cut.

It became clear early on that British companies were not the only ones to take advantage of the consumer boom. Even such basics as cement had to be brought in from abroad and, while the building supplies industry worked flat out, it could not keep up with rocketing demand. The problem with consumer goodies – from washing machines to video recorders – was worse. Far Eastern companies were busy setting up assembly plants, but the only big British washing machine maker was GEC's Hotpoint, which took over TI Creda in 1987. These were not enough to stop the trade deficit on manufactured goods rising, from £3 billion in 1985 to £5.8 billion the next year and £7.3 billion in 1987. By 1987, imports satisfied 20 per cent of British consumer demand, having climbed steadily from 11 per cent in 1970. That was hardly surprising – the fastest growing categories of expenditure were television, video, cars and clothes: all areas where imports were taking over.

For the moment at least, this was not causing too much concern. Inflation, the government's ultimate measure of success, stayed obediently low. Having dipped down to 3 per cent in 1983, it briefly surged above 5 per cent in 1985 before dropping back to a modest 3.7 per cent by the end of 1987. Falling commodity prices were crucial: the oil price collapsed in 1986 and overall raw materials prices fell by 30 per cent relative to manufactured prices between 1984 and 1987. Interest rates moved up when the pound was low in 1985, but had been falling since then. The chancellor, worried that the consumer spending boom could reawaken inflation, had put them up in August 1987, but after the crash started to ratchet them down again to make quite sure the economy did not tip into recession. As America continued to boom, world trade increased, and basic industries – notably steel – basked in a friendly new climate.

It was easy to claim that economic improvements were no more than cyclical, and that industry would in due course plunge back into the black hole whence it had come. Statistics could be used to prove or disprove that. The Labour Party pointed out, correctly, that relative to other countries British unit labour costs had never been so low under the Tories as they had been in 1978. The Tories could point out, correctly, that unit labour costs had been falling in real terms since 1982. Critics could say that capital investment was running at a lower level than it had been in 1979; supporters could then claim that that was irrelevant, as capital was now being used more efficiently. On some measures, British industry would be overhauling the Germans within a decade; on others, the gap was getting even greater. It was all rather inconclusive.

But there were some simple and indisputable points that could be made. First, the British share of world manufactured trade had stopped falling in 1981, and was now starting to move gradually upwards. That reversed a thirty-year trend. Second, profits were expanding rapidly: return on capital employed had recovered from a nadir of 3 per cent in 1981 to 9 per cent in 1987, a rate previously achieved in 1972. Third, wages were also rising fast, by 7.5 per cent in 1985 and 1986, and by 8.25 per cent in 1987. And fourth, manufacturing was never again going to be the major employer in the country: in 1986, as total unemployment reached a plateau, another 142,000 people were thrown out of work in the factories.

The first point was a particularly important indicator because it showed that somehow, whether through better pricing, quality, design, marketing or whatever, British companies were starting to fight back against foreign rivals. Relative performance was what decided success in a free market, and that had clearly been improving. Rising profitability could have come from easier markets, better performance, or both; it would take a recession to find out which. And higher wages were largely a function of better profits. Government ministers ceaselessly pointed out that if wage rises had been suppressed, British industry would be even more competitive. The trouble was, those ministers also praised the virtues of profit sharing, and profits were rising fast. Sometimes, the workforce was formally given a slice of the profits; more often, managers were not inclined to fight too hard for low settlements. They may have noted that the Japanese in Britain preferred to fight costs in areas other than pay – the wages they paid, always linked to some sort of performance measure, tended to be generous.

The fall in manufacturing employment needs some amplification. From 1981 to 1985, the number of small companies in business services grew from 43,000 to 64,000; and in 1986, a survey by the *Financial Times* showed that 35 per cent of companies were contracting out work that had previously been done in-house. These figures fit together. If a manufacturer started employing an external data processing bureau instead of its own staff, manufacturing employment would appear, misleadingly, to have fallen.

Rank Xerox was one of the first companies to use the services of previous employees on a freelance basis, with its Xanadu scheme. The trend towards contracting out everything from cleaning to research and development, which paralleled the growth in temporary workers, had its basis in a mix of common sense and business faddery: it saved on overheads, it allowed flexibility when business turned down, and it obeyed the dictum that companies should stick to their knitting.

Sometimes, it was difficult for observers to sort out companies that had really got to grips with their problems from those that had risen up on the back of booming demand, or were just good at wheelerdealering in the bull market. A company in the building or consumer trades would have to try hard not to have its profits rise. Just as the toy industry collapsed with consumer spending in the early 1980s, so it was now recovering. John Waddington's games division was going from strength to strength and, at the end of 1986, Hornby was floated on the Unlisted Securities Market. It had been engulfed in the collapse of Dunbee-Combex-Marx in 1980, but had been rescued by a management buyout. The renamed Hornby Hobbies recovered from a £1.2 million loss in the year after the buyout to make almost £1 million profit in 1985. Hornby trains and Scalextric racing cars, an integral part of so many youths, had been saved and – more surprisingly – were still British.

But the healthy market could also disguise surprising stories of success, where a combination of recession and predator-induced regeneration had led to turnaround. What was remarkable about these was they all seemed to be based on the same formula. Managements everywhere were looking for niches, moving up market, flattening their structures, pushing authority downwards, introducing tight financial controls ... and so on.

Companies that no one had heard of suddenly emerged like butterflies from their chrysalises, having received a dose of the magic formula. Many of the most successful companies were obscure not just because they did not make consumer products, but because they did not make complete products at all. The industries where "global niches" were easiest to find were those that did not make things, but that made bits of things and, even as the British-owned car industry disappeared into oblivion, the motor industry as a whole was fighting back. Companies like Tallent, or on a larger scale Lucas or GKN, were carving out niches for themselves. GKN dominated the world supply of constant velocity joints – the heart of a front wheel drive car. Lucas became one of the world leaders in its chosen areas, including braking systems and guided missiles, while Dunlop's aviation division (now owned by BTR) was cutting into the sophisticated aircraft brake business.

Perhaps the least known butterfly to flutter forth was Cookson Group which, in 1987, turned over more than £800 million (making it Britain's 134th biggest company), compared with £280 million in 1981. It had a wonderfully obscure niche product. With ICI, it owned Tioxide, one of four makers of titanium dioxide, the chemical that makes white paint. New plants were so expensive to build that competition was locked out of the market, and as the DIY and building markets flourished in the mid-1980s, Cookson's only worry was how its other activities could stop Tioxide dominating its profit figures. In 1982, it had thrown off its old name, Lead Industries Group, and set off on an acquisition trail buying up companies in a series of related niches. Each provided specialist materials for industry: in America, for example, about half its $1 billion sales went to the electronics industry, providing a bewildering range of materials for printed circuit boards and the like.

Avon Rubber – better known, but not much – moved itself sharply up market. At the beginning of the decade, half its profits came from run-of-the-mill tyres, where it was competing against the might of the giants such as Michelin. By 1988, tyres as a whole accounted for 25 per cent of sales, and half of these were specialist, going to companies such as Rolls-Royce; the balance came from specialist industrial chemicals. At the same time, Avon was introducing a full set of Japanese-style productivity improvements. In 1982, it lost £6 million; in 1987, it made £11 million.

Folkes Group – the old John Folkes Hefo – took its time struggling out of the chrysalis, but it got there in the end. Since Constantine Folkes had taken over as chairman in 1981, he had been remodelling the family firm. Although it was quoted on the stock exchange, a controlling interest was firmly in the hands of the family. That meant it was a company that the City barely bothered to look at, but when it did it was to cluck with displeasure. There was nothing it hated more than a company that could not be taken over: it must be badly run.

Folkes was not. In 1981, it had lost £724,000, and only its property business had kept the figures in the black, just, for the next two years. Although he had been busy shedding businesses, cutting costs and shaking up management, Con Folkes hung on to the heavy forging business, and for this he was chastised. In 1986, when heavy forging was coming back to life, the *Financial Times* complained that the group's restructuring had been "very, very slow". The engineering loss-makers should, it implied, have been shut down years ago.

In fact, the apparent sloth had been quite deliberate. "We took our time so that we could sell off companies rather than close them down," Folkes says. By 1987, there were ten companies left. They were something of a ragbag, ranging from foundries to a fitted kitchen maker, but the group's structure was familiar enough: autonomous units, tight financial control, defined niches with a decent market share. Profit had climbed back up to a modest £2.6 million, and the chairman – now an elderly 34 and with his first grey hairs – decided it was time to buy himself a Rolls-Royce again. He went for one of Peter Ward's Bentley Turbo Rs.

The month before the stock market crash, Folkes announced it was spending £2.5 million at BarBright, a manufacturer of bright drawn steel and its biggest subsidiary. Eighteen months later, with Britain's balance of trade tipped another £1 million in the wrong direction (the machines were Italian), the company announced that it could now produce the best quality drawn steel in the world and was exporting it throughout Europe. Con Folkes had done what a good predator would have done – but there had been a lot less bloodshed on the way.

In 1985, Ronny Utiger must have dreamt of having a Folkes-type ownership structure. Recently appointed chairman of TI Group, he found himself staring down the barrel of a bid threat from the upstart predator Evered. He was not in a strong position to resist. When the recession hit, TI had been a ragbag of engineering businesses, grouping components companies alongside Raleigh bicycles, Creda washing machines and British Aluminium. In 1981, it went into the red, and British Aluminium was sold off for a pittance. But in 1984, losses at Raleigh wounded the group terribly, and predators started circling overhead; that was when the Abdullah brothers started to build up their 20 per cent stake.

Utiger knew TI needed a first-rate chief executive, and searched for one frantically. In 1986, he appointed Chris Lewinton, an engineer who had spent most of his career with Wilkinson Sword, twelve years of it running its American businesses. Lewinton allowed himself 100 days to size up the problems, then hurled the group into an upheaval unmatched by any big company, save possibly Lucas. His first job was to convince the City that TI was going to change; like Gareth Davies at Glynwed, he gave it a promise: that the group would earn 10 per cent return on sales, against the current 5.6 per cent. Then he started to reshape the group. His aim was to shed all the consumer brands, and to concentrate on specialist engineering, digging out a row of "global niches".

In 1987 alone, eight companies were sold for £260 million, and three were bought for £330 million. Out went companies like Raleigh, Creda, Parkray and Russell Hobbs, which were earning 3 per cent; in came a series of little-known companies with leading positions around the world. Their average return was 10 per cent. Building on Midlands-based Fulton (TI), which supplied 70 per cent of all small-diameter tubes for brake and fuel lines in Britain, Lewinton bought the European operations of Armco of the USA, and another American company, Bundy. These two acquisitions, costing £112 million, gave him 40 per cent of the world market for small-diameter tubes. The same trick was performed with mechanical seals: by buying John Crane, an American company, TI won a 30 per cent share of the world market. John Crane International, as the seals group was called, had 4,000 staff in twenty-two factories. The Abdullahs saw that Lewinton was doing what they would have done and sold their stake.

Inevitably, manufacturing jobs in Britain were sacrificed for the international goal. When Lewinton arrived, 55 per cent of TI's sales came from the UK, 21 per cent from Europe and 15 per cent from North America. After two years, the figures were 25, 25 and 40. In 1988, TI earned 8.9 per cent on its sales, and the City loved it. But it had been a close thing.

The process of internationalisation accelerated in 1987. British companies turned their acquisitive eyes towards America and spent $32 billion there during the year, twice what they had in 1986. There was hardly a big British company that did not at least go window shopping across the Atlantic. Most of the boldest deals were in the service sector – WPP buying J Walter Thomson, Blue Arrow buying Manpower – but Michael Montague, the cultured former chairman of the English Tourist Board, held up the flag for manufacturing. His heater company Valor tripled in size by buying both NuTone, whose base product was doorbells, and the lock company Yale. When he first thought of Yale – it was one of the few true "global brands" – he did not even know it was American. But investigating further, he found it was being milked for cash by its Canadian owners, who had to pay off junk bond debt, and was in dire need of rescue. NuTone was in the same boat – so he threw them both a lifeline.

The biggest manufacturing deal of all was Hanson's $1.7 billion takeover of Kidde, conducted by Sir Gordon White for the usual opportunistic reasons. A more typical takeover was ICI's strategic purchase of Stauffer Chemicals, also for $1.7 billion; by now it was a giant in its own right in the USA. But the record of British acquisitions in America had not been good, and this latest spree had its victims too. Within two years, Blue Arrow's Manpower deal, as well as Ferranti's £425 million purchase of International Signal and Control, had turned sour. Too often the British were foiled by the language into thinking they understood the business culture.

In absolute figures, the stock market crash hardly depressed takeover activity; there was barely a month's lull before bids started popping up again. By the end of the year two, worth more than £2 billion apiece, had been declared: BP's for Britoil, which went through, and Barker & Dobson's for Dee Corporation which did not. But the character of the takeover boom was transformed. The companies with the power now were those that had been thrifty, saving up their cash, keeping down their debt, and keeping their balance sheets as strong as they could. The losers were those that had relied too much on their glamorous share rating – and had not matched their dealmaking skills with good management. Some of them found themselves turning from predator to victim.

John Crowther was in the first rank of predators. Trevor Barker, an accountant from Teesside who had already made one fortune in the travel trade, bought Crowther as a near bankrupt shell in 1981. It was a textile company, and Barker used it as a vehicle to hoover up other textile and carpet businesses in the bad years of the early 1980s. As sterling dropped, the company became increasingly profitable, and Barker went on buying: in 1986 alone, he bought fourteen companies. In the five years to 1987, Crowther grew by an annual average of more than 200 per cent, ending up with sales of almost £358 million and profits of £25 million: by then it was the 247th biggest company in Britain, with a growth record matching the most aggressive of the conglomerates. But the City was getting suspicious. It thought Barker had paid too much for some of the businesses and that he did not have the management skill to make the most of his acquisitions. What was worse, some subsidiaries were stubbornly refusing to make money.

When the Crash came, John Crowther was hit harder than most and its share price stayed low. Trevor Barker decided that he had taken the company as far as it could go, and when in April 1988 John Ashcroft of Coloroll made an offer, he agreed to sell out. He then sat back and watched while Graham Rudd (Nigel's big brother and head of a new conglomerate called Thomas Robinson) made a counter-bid. After a tussle, Coloroll won and paid £208 million for the company. Even after Coloroll had sold the textile business to a management team, it was still the biggest home product company in Britain: in 1988, it turned over £565 million, and made a profit of £56 million. That made it Britain's 99th biggest manufacturer.

This takeover was interesting, the first time a predator had eaten a predator, and it looked as though this might turn into a trend. In mid-1988, Williams took a threatening stake in the newly created Yale and Valor: the City was doubtful of Montague's ability to control such a far-flung empire, and internal management strife had not comforted it. But Williams went elsewhere, and Yale and Valor survived. It would take another recession before there was a real shake-up among the predators.

The sharp drop in share prices encouraged the emergence of another sort of business creature, half predator, half entrepreneur, perhaps best called an instant tycoon. His technique was a new and useful one, known as a management buy-in.

On the day of the stock market crash, Alan Bowkett could have exercised his stock options for the first time. He was the highflying 36-year-old managing director of BET's subsidiary Boulton & Paul, and had already mentally spent the £100,000 that he reckoned would be coming to him. He had to make do with half that but, rather than mooning around, he went out and bought a company for £74 million. It was the biggest management buy-in so far.

The buy-in was an obvious development from the management buy-out. It was where a company, or part of a company, needed new management, but there was no one able or willing to take on the job from inside. Managers in the complementary position – wanting to run a company but not working in one that was available – would raise the money to buy the company and then, using their skill and judgment, improve its performance. It filled an obvious gap in the process of industrial restructuring and allowed would-be millionaires to hit the ground running, rather than having to build up from a small base. The catalyst in most management buy-outs was a venture capitalist: 3i organised its "break out" programme specifically to tempt managers away from large firms.

Bowkett was the very model of a modern manufacturer. He came from a mining family that lived in a village near Newark, in Nottinghamshire; he went to the local grammar school and on to University College, London, where he read economics. He then went on to take an MBA course at the London Business School, and joined the service group Lex, where he became corporate planning manager for the industrial division. He was in New York during the British recession, but returned to take a job with BET, a conglomerate that was nervously watching the prowlings of Owen Green: there were sighs of relief when he finally pounced on Thomas Tilling. But the fright had had its effect and BET's new chairman, Nicholas Wills, started to use bright young men like Bowkett to shake the company up. Working in the centre, he spent his time buying and selling companies; he was rewarded with his own company, Boulton & Paul. Once manufacturer of the wartime Defiant nightfighter, it was now the BET's second biggest company and a somewhat fatigued maker of building products and steel fabrications.

He put his business school skills into practice, analysing market share and competitiveness, applying a clinical professionalism to the business. He also learnt what business school had not taught him – about working with people, how to judge the balance between being too familiar and too distant, and how to break down the barriers between the shop floor and management. He was successful, and could have been managing director of BET in his early 40s – had he not had a burning desire to be wealthy. "If I made £5 million out of a venture," he says, "the vast majority would go into an investment portfolio – but I would buy a house in Tuscany, pay off the mortgage, take two decent holidays a year and not have to worry."

The next move was obvious. "I met Greg Hutchings for lunch," he says. "I thought, he's a smart guy, but I could do that." With Gavin Morris, a former colleague and fellow MBA, Bowkett spent his weekends sifting through balance sheets, looking for companies that were turning their stock over and collecting their debts too slowly – sure signs of sloppy management. His brother-in-law was a director of Citicorp Venture Capital and, with his promise of backing, he made offers for several companies. None came to anything.

If the ill wind of the stock market had swept a few bedrooms away from Bowkett's Tuscan villa, it also blew in a spectacular opportunity. One of the companies he was eyeing up was the bearings division of Ransomes, Hoffman & Pollard. RHP was a creation of the Industrial Reorganisation Corporation, constructed out of Britain's biggest ball-bearing companies in 1969 to take on giants such as SKF of Sweden. But rather than concentrate on getting its bearings business right, it had diversified into electrical components and fire prevention businesses and, by 1987, had decided that bearings were no longer essential to its existence. The obvious buyer, SKF, would have faced a Monopolies Commission reference. Other companies could only pay with shares and that, post-Crash, was impossible. RHP appealed to Bowkett: he remembered how his mother had been sent to work at the Newark factory in the war and how, overawed by the huge gates, she had run away to sea to become a Wren, and met his father. It seemed the right company to buy.

November 1987 was not the easiest time to raise £73.5 million but, with the help of Citicorp, which arranged a £10.25 million slug of venture capital, he was offered a £44 million loan by Standard Chartered. The loan was split, or syndicated, between the clearing banks. Bowkett, Morris and Boulton & Paul's manufacturing director, Roy Hammond, then went along to the National Westminster to ask for a loan to cover their share and to allow twenty other senior managers to buy shares; it was approved in three minutes. "I don't think we would have got that same answer ten years ago," Bowkett says.

Although interest rates were then low, he was canny enough to see that they might rise: he arranged to "swap" most of the debt into a fixed rate loan.

As part of the deal, he was allowed to keep the RHP name; the rump of the group changed its name to Pilgrim House and in due course disappeared into the jaws of Williams. Although the company turned over £95 million and was making a £10 million annual profit, Bowkett had no illusions about its real state. A ballbearing was the nearest thing to a widget there is, but they were not easy to make well and cheaply. When he took over, Bowkett says, "the cost base was horrendous".

The seventeen-acre factory had a long way to go. Built in 1908, it looked as a factory should: sharply pitched roofs, glass-filled on one side; plenty of antediluvian machinery in neat rows; in the turning shop forty-two machines, mostly American, running twenty-four hours a day. Some of the buildings had been bombed and rebuilt. "Unfortunately the RAF did a better job on German factories than the Germans did on ours," said a manager. Many of the machines were forty years old, although statistical process control had been introduced under pressure from Ford; there was talk of bringing in just-in-time one day. Vast heat treatment machines, with "British Furnaces" stamped on the side, heated the bearings at 860 degrees, then drenched them in oil to harden them. The process had not changed for thirty years but, at the insistence of the buyers, accuracy was having to be constantly improved; that was increasingly difficult.

Bowkett was starting a process that should have begun years before. He closed down a factory at Chelmsford, and had to work hard to dampen down industrial unrest; his approach was typical of a new-style manager – a ruthless insistence that unprocedural strikes would not be tolerated, combined with attempts to break down the management/shop floor gap. He introduced cascading team briefings and, as his own contribution to the art of management by walking around, wore bright red braces "so people can't say they didn't see me". But, he says, "you can try all the modern techniques, but it doesn't happen overnight."

Bowkett knew that he could never take on SKF or the other bearings giants on all fronts, but hoped to carve out niches – in bearings for aircraft, for instance. He put great faith in his managers: if they performed, according to the financing arrangements, they could turn £2,000 into £100,000 through the share scheme. But much of the burden would fall on him, and this led to the £73.5 million question. "I find it very difficult changing from dealing to managing," he admitted, "I still get an immense thrill from a deal. Now I have three years of balls-aching hard work to get from £100 million to £200 million ... I think the jury is still out on whether dealmakers *can* turn into managers."

For the predators 1987 and 1988 were a time to show how astute they were. That meant showing that they could do more than just deal. At the beginning of 1986, Williams Holdings was a rather wobbly small conglomerate. By the end of 1988, it was the second largest paint maker in Britain and a member of the exclusive club of blue chip companies that made up the *Financial Times*–Stock Exchange 100 index. The three other WETS, Evered, Tomkins and Suter – with whom it was level-pegging in 1986 – were left spluttering in its wake.

Not that everything had gone smoothly for Rudd and McGowan. In early 1986, they had tried to triple the size of Williams by buying a Midlands plastics and metals company, McKechnie, for £140 million. McKechnie thwarted them by buying another Midlands engineer, Newman Tonks: that was what the Americans called a "poison pill" defence. Then in April 1987, they offered £570 million for Norcros; it was half as big again as Williams and in the 1960s had been just as glamorous. Now, it was working hard to rebuild its fortunes as a building products and packaging group, and just managed to convince enough fund managers that it deserved its independence: Williams' bid failed by 2 per cent. Rudd and McGowan decided that they would keep away from hostile bids in the future.

Agreed takeovers more than made up for these failures. In 1986, Williams bought Fairey Engineering (once an NEB lame duck, and the world leader in military bridges), Duport, maker of Swish curtain rails, and the conglomerate London and Midland Industrials, which had a tempting collection of American businesses. It also sold twenty-one businesses, including Ley's Foundries. The group was less of a pure conglomerate now; it had identifiable legs: branded building products, specialist engineering and, incongruously but profitably, up-market car dealerships.

Since 1984, Rudd and McGowan's great fear had been that the bull market would end and that they would be left with too much debt and too few profitable businesses. The J & HB Jackson deal had answered the debt problem; now they were concentrating on "quality" companies. They knew that to go on growing when the bull market finally died, they would have to have "critical mass", to dominate markets in exactly the same way as Lucas, GKN or TI did. But the world was not their market: brands were powerful only where they were advertised and that, for the vast majority of them, meant in one country. Building on Rawlplug, they added Swish and then in July 1987, they paid £260 million for the Crown Paints and Polycell businesses of Reed. A management buy-out team was ready to take Crown out of Reed when Williams swooped. "If a management buy-out is going for a business we want, we would always overbid," says Rudd. "It's a good endorsement." The leader of the MBO team, Paul Lever, was put in charge of the new paint business.

By December 1987, the hit squad had doubled the profitability of Polycell, and it was in that month that Williams paid Hoechst £130 million for Berger, Jenson & Nicholson, which made Magicote and Brolac paint. To finance that, Rudd and McGowan asked seven City institutions for £100 million. It was the first big share placing since the October Crash, but such were the two men's persuasive powers, they had agreement within a day. Their argument was that the deal would give Williams 25 per cent of the British paint market, not far behind ICI's 34 per cent, as well as three big brands to pitch against ICI's Dulux.

Williams applied its usual combination of cost cutting and investment. After analysis by the hit squad, Berger's Australian operations were sold to ICI and its Bristol factory was closed down. Paint production was concentrated from four factories into two. The main one was at Darwen, tucked away in the rainswept Lancashire valleys north of Manchester. In among the old cotton mills (one, owned by Coats Viyella, was still working), Reed had created one of the most modern paint plants in Europe. Williams spent another £11 million on it, doubling production and moving on to triple shift running. One million litres of water-based paint and 400,000 litres of oil paint were produced every week; there was hardly a soul in sight.

The other side of the Williams formula was marketing, where spending on the paint division was doubled: in the first two years of its ownership of Crown, the group spent £15 million on television advertisements alone. Paul Lever decided he was glad that his management buy-out had not been completed. "On reflection, an MBO would have been very difficult," he says. "The company would have been highly leveraged and we would have to run it for the cash. What is good for the management may not be good for the business in the long term."

In 1988, as house prices peaked, Williams started to swing away from building and DIY products, which were bringing in 60 per cent of profits. It bought Smallbone, the upmarket fitted kitchen maker in August (Graham Clark, who founded it in 1980, found himself £7 million richer) but then careered off in a completely different direction by paying £330 million for Pilgrim House, previously RHP, in October. It made electrical and electronic equipment in Britain, Europe and the USA, and was trying to buy Kidde's fire protection division from Hanson. British predators were still bestriding the Atlantic, and making the occasional deal with each other.

At the beginning of 1986, the market capitalisation or stock market value of Williams was about the same as that of Evered, Tomkins or Suter. Two years later, it was about the same as those three's put together, and was the only conglomerate that looked as though it could achieve the might of BTR or Hanson. In 1988, its £820 million sales made it the 68th biggest manufacturing company in Britain, Hanson was the third, BTR the eighth, with Tomkins, Evered and Suter at 147, 165 and 190. Of the predators that started up during the decade, only the "nice" one, Hillsdown, had grown faster than Williams. David Thompson retired in 1987, but Harry Solomon kept up the pace, buying, among other companies Canada's largest flour producer. In 1988, Hillsdown was Britain's fourteenth largest manufacturer by turnover, with sales of £3.5 billion – although the comparatively low margin nature of the food business was reflected in profits of £150 million, not far ahead of Williams' £115 million.

David Abell of Suter, which in 1985 had appeared to be the strongest of the mini-conglomerates, was known as a dealer. In 1987 a television programme raised questions over his share purchases, and he was sent slithering back to the beginning of the board. The Abdullah brothers at Evered made one big acquisition, of Rudd and McGowan's old company, London and Northern. They then sold off all except the quarrying interests and transformed the company into a straightforward building products company; in 1989, they left the board. Greg Hutchings at Tomkins was building a conglomerate, but his growth was limited by an unwillingness to issue shares to fund his purchases. On the great American shopping trip in 1987, he had come back with an intriguing prize, though: the gunmaker, Smith & Wesson. None of these predators could be called failures, but they lacked the combination of dealing skill, management and boldness that marked off the Williams men.

Meanwhile, a select band of veterans of the recession had dusted themselves down and were doing their best to catch up with the real predators. These were the people who had come within an ace of annihilation in the early 1980s, but were now determined to do more than just survive. They would never, thanks to the ending of the bull market, catch up with Williams – but they might well overhaul some of the more laggardly predators.

*

After his triumphant seizure of RFD, the bullnecked figure of Wardle Storey's Brian Taylor was watched expectantly by the City, and with some trepidation by potential targets. He made no secret of his desire to buy, and was now as astute an analyst of annual reports as any predator. He had a rule of thumb: "If a company collects debts too late and pays them too early, it is a bad company." But he went further than most in finding out what was really going on, by using an "industrial research" company called LEK Consultants. It would ring up managers and, by asking them apparently innocent questions, construct a detailed picture of the firm. "You would be amazed at the large companies whose acolytes speak recklessly," he says.

But there were some things even Taylor could not prepare for – one of them was the City's flight from hostile bids. In February 1987, just after BTR had shied away from Pilkington, Wardle Storeys bid £44 million for Chamberlain Phipps, a manufacturer of shoe components and adhesives. Two days after Williams lost its bid for Norcros, Taylor discovered he too had been defeated; "sentiment" in the City had turned. So he settled back to his old ways, keeping prices up and costs down and pumping return on sales up from 8 per cent in 1984 to a remarkable 22 per cent in 1988 (the average for manufacturing industry was about 9 per cent). But although his research team was developing specialist survival equipment for aircraft and ships he knew that the potential for the core PVC business was limited. He needed to diversify.

In November 1988, he made another bid, this time attacking a motor components group, Armstrong Equipment. Armstrong was, Taylor believed, one company that still had plenty of scope for improvement. Two months later, he dropped the bid and declared that he would never again attempt a hostile takeover. Just as he had constantly chased increased profits rather than volume, he refused to pay a penny more than he thought a company was worth. That was why, in the five years from 1984 to 1988, his company's profit increased more than fivefold, to £16.5 million, while sales had not even doubled – they were £76 million in 1988. "I would rather be remembered from increasing my earnings per share than for creating a huge business," he says. There was a point, though, at which the pursuit of extra profitability would inevitably cut into sales: that was what Taylor had to be careful of.

Tim Hearley, the dapper ex-stockbroker with an office facing out over London's Cavendish Square, was a very different type. But his company, CH Industrials, was one of the few that could keep pace with Wardle Storeys. Its turnover rose from £13.5 million in 1982 to

## Crash? What Crash?

£110 million in 1988, and profit from £400,000 to £9.5 million. It provided yet another variation on the manager/dealmaker theme, with an unusual mix of conglomerate-building and high technology.

CH's acquisitions were frequent, usually small, and usually friendly: its aim was to build up a group that balanced cash generators against cash users, low tech against high tech, with "synergy", sometimes vague, between the companies. It bought an office furniture maker and shopfitter because CH already made foam and chair covers; a company that made train windows and doors because Aston Martin Tickford could design the carriages; and Siegel & Stockman, the doyen of shop mannequin makers, because it could open doors for other parts of the group that supplied shopfitting products. Also, of course, each company had a tight niche, and was kept under strict financial control from the centre.

The company reinforced its niches both with design expertise, and with research and development. Some of its subsidiaries operated at a level of technology that would have made Lord Hanson twitch. Aston Martin Tickford developed a new racing engine, did General Motors' production engineering for the Buick Reatta and built the Ford Cosworth 500. In 1988, CH bought Motor Panels, Europe's biggest cab manufacturer, and maker of special bodies (in 1959, it had built Donald Campbell's world record breaking car Bluebird). That meant Hearley could plan the next step: to set up a factory that produced limited runs, up to 10,000 cars, of "specials". It would be the only company in Britain that could do that.

In 1987, he made his first big acquisition, buying Gripperrods, the definitely low-tech carpet-fixing company, for £30 million. The aim was to generate cash for the automotive sector, and also to counter the group's dependence on it. At the same time, he was trying to solve a problem that was starting to affect subcontractors everywhere, particularly in the automotive business.

The problem was this: as the subcontracting business had been remoulded, with fewer companies being given more business, they were expected by their clients to play an ever greater role in design and development. With Aston Martin Tickford, this was not a problem – it was a research-based company. But what about the sunroof maker Tudor Webasto, and the CH companies that made window frames, seats and body trim? How could these minnows keep up with the Germans, who had most of the largest automotive suppliers in Europe? It was a particularly sticky problem for the British, who had been so squeezed by the recession that any development they might have been involved in had gone by the board. In some areas, they were ten years behind the Germans.

Hearley reckoned that the only way to keep up with the giants was to join them. He set up six joint ventures with German companies, and one each with a Swedish and American partner. He had pioneered this sort of relationship with Webasto of Germany, and was now repeating the formula – CH took the technology, kept management control, and in exchange offered the partner expanded sales and profits from the UK market. Although this restricted its ability to export, it made it much easier for CH to become the market leader in the UK; by 1988, it had 60 per cent of the sunroof market.

On April 13, the Swiss company Suchard gave notice that a new type of predator had arrived in Britain; it was given various names – "1992" was one, and "international brands" another.

Kenneth Dixon, the mild-mannered chairman of Rowntree of York, was visiting a subsidiary in South Africa when he heard that Suchard had just gobbled up another 15 per cent of his company in a dawn raid. He was not wholly surprised: speculation that Rowntree might be pounced upon had appeared regularly in the press for the previous four years, and he had already been approached by various companies, including Suchard, about a link-up. His answer each time had been, no thanks, we're quite all right on our own. But, as he flew north out of autumn into spring, he wondered if he should perhaps have listened a little more carefully.

As a device to persuade fund managers to yield up their shareholdings, Suchard had promised that it would not make a full bid for a year, and that it would not pay more for shares than the £6.30 it was now offering (the previous night's price had been £4.77). Both conditions lapsed if another company made a bid and that, Dixon feared, was exactly what would happen. Arriving back in England, he was given a brush-up lesson on television technique by his public relations firm then, aided by David Challen of Schroders (hero of the Pilkington defence), he got down to pouring scorn on Suchard.

Dixon knew quite well that Rowntree was vulnerable. Although he had squeezed costs and even made redundancies during his eight-year chairmanship, he respected the firm's Quaker tradition too much to act with anything approaching ruthlessness. After a bad year in 1980, when sweet sales were hit by the 1979 VAT rise, Rowntree had regained its even keel, living off a fairly steady sweet market. It had also started to diversify into snacks, buying companies in America, and had worked hard to build up the brand image of such goodies as Lion

## CRASH? WHAT CRASH?

Bar, After Eight and Polo in continental Europe, where it had four factories. By 1987, Dixon was running a fairly successful, if not sparkling, company: it made a £12 million profit on sales of £1.4 billion that year.

It was those European brands that made Suchard salivate. It had Toblerone, but that was all; it thought the British, who were so brilliant at consumer marketing, badly underestimated the international power of their brands. Unfortunately for Dixon, so did Nestlé; the giant from Vevey, which had profits as big as Rowntree's sales, swooped in with a full bid on April 26. There was already a certain rivalry between the two Swiss companies, and no one was surprised by the move. But they were astonished by the price: Nestlé was offering £8.90 a share, or £2.1 billion for the whole company. This was way beyond what the fund managers thought it was worth, but they could see there was more to come. A full-scale battle was in the offing, and that could only push the price up further.

Suchard, which knew it could not match Nestlé's resources, played a secondary role in the fight. It was Dixon and Challen's fierce rearguard action that did wonders for the share price. What the industry minister, Kenneth Clarke, dismissively called a "northern populist revolution" was whipped up in York; boxes of chocolates were sent to MPs, and anyone with any influence was lobbied to try to get the bid referred to the Monopolies Commission. The argument was not about monopoly as such – Nestlé's chocolate division had a tiny share of the British market – rather it was based on reciprocity. British companies could not buy Swiss companies, because of their restrictive share structures, so why should Swiss companies be allowed to buy British ones?

The Office of Fair Trading was not convinced. On May 25, Lord Young, the trade and industry secretary, announced that he would not refer the bid, and from then on all the Rowntree team could do was to negotiate the best possible price. It did well. On June 23, Nestlé said it had bought Rowntree for £10.75 a share. There was anguish in York, but as half the workers had shares or share options, it was an anguish tempered by the promise of hard cash.

The takeover was important, because it gave several signals as to the way that the British consumer industry was going. First, it put a value on brands. The final price Nestlé paid was two and a quarter times the pre-bid share price. Had Nestlé been a British predator, that could only have meant one thing – that it saw a potential for huge extra profits, by squeezing, by investing or whatever. But Nestlé did none of those things: it thought that Rowntree was worth £10.75 a share as it was, simply because it had those powerful brand names. People in industry had understood the value in brands for a long time, but now they had an opportunity to put a figure on them, and not be laughed at by the City. Ranks Hovis McDougall brought in a specialist company to work out the value of Bisto, Saxa and its colleagues. Having studied factors such as their age and market position, it decided they were worth an impressive £678 million, and that amount was added to RHM's assets. Throughout industry, the share price of brand-owners moved sharply upwards.

More important, the deal had made it clear that Nestlé and Suchard believed that the consumer goods market was becoming ever more international. In 1986, United Biscuits had taken the same line when it tried to merge with Imperial Group. Then, the City institutions had chosen to ignore its argument and had sold out to Hanson. Perhaps UB had been right after all? It was an argument that had more potency in the summer of 1988, for Lord Young's "1992" campaign was in full swing, and people who failed to talk in terms of Europe were liable to be roundly chastised.

There was a belief that Nestlé was buying Rowntree to establish itself inside the Common Market before great barriers were thrown up after 1992. That, like many things to do with the European single market, was a misconception. For one thing, Nestlé was already a giant in its own right inside the EC; for another, the single market was unlikely to stop companies selling into the community – indeed, by harmonising standards, it was likely to make it easier. So it was not surprising that the Rowntree takeover did not presage an invasion of foreign predators. The people who took "1992" most seriously, and had most to fear from the erection of new barriers, were the Japanese, but they preferred "greenfield" investment.

In Britain itself, companies ignored the Single European Market campaign at their peril: to be go-go, it was imperative to make the right noises. A handful of large companies – ICI, Lucas, GKN among them – were already well-established on the Continent, and many firms restructured their marketing divisions into a more euro-friendly way. Amstrad was particularly successful in France and Spain, where it built up its own distributor network. But, although there was an upturn in acquisitions in Europe, they were dwarfed by those in the States: in 1988, British companies spent thirteen times more in America than they did on the Continent. There was some interest in buying companies in low-cost Spain, or in Germany, where private firms set up after the war were starting to come on to the market. But it was no coincidence that the country where most companies looked first was the one that exporters used as their kindergarten – the culturally friendly Netherlands. Hillsdown was one of the companies that bought up Dutch distributors. Whereas British companies had not been cautious enough in America, being fooled by the language into underestimating the cultural gap, they were terrified of the Continent. "It's no accident that we haven't bought anything straight into Europe," says Nigel Rudd of Williams. "You have to be aware that you must have a national on your side – a Williams man who will represent your interests." He was right, but such bi-cultural creatures were hard to find.

In any case, takeovers were essentially an Anglo-Saxon phenomenon. If British companies were going to treat Europe as their home market, as the 1992 campaign urged them to, they could not impose their predatory ways willy-nilly. Little understood and rather frightening concepts such as joint ventures would have to be considered, and even adopted. Nestlé had suggested that it take a 25 per cent share of Rowntree in 1987, which would have protected the York company from full takeover. Sooner or later, the ghost of Pirelli-Dunlop would have to be faced again.

20

# *Family silver*

In March 1986, even as the government's attempts to sell bits of British Leyland to the Americans were collapsing, Graham Day was quietly signing away the most profitable part of British Shipbuilders, its warship yards. The Vickers submarine yard at Barrow-in-Furness was bundled together with Cammell Laird at Birkenhead and sold to a management team headed by the Vickers' chairman, David Nicholson, and its chief executive, the nuclear scientist Dr Rodney Leach. Trafalgar House, the conglomerate that owned Cunard, outbid the team, but with the Land Rover affair still bubbling, the trade and industry secretary Paul Channon judged it expedient to let the team go ahead with their £100 million buy-out.

Immediately, it embarked on a programme of popular capitalism more successful than any of the government's mass privatisations. The staff and local people fell over themselves to subscribe £7 million for shares in the company: 11,500 workers, more than 80 per cent of the total, became shareholders. In July, VSEL Consortium, as the company was called, was floated: the promise of a Trident nuclear submarine to build made the group something of a blue chip. Solidly working-class Barrow now found itself with a higher proportion of shareholders than any other town. In 1983, it had voted in a Conservative MP for the first time in many years; his seat was now safer than ever.

The Conservatives had two privatisation programmes. One, which got under way with the flotation of British Telecom, affected monopoly suppliers of services; it was ideological and controversial. The other, which included all the manufacturers, unravelled the industrial policy of past Labour governments, but was pushed through with hardly a squeak of protest about the principle (even if the detail stirred up several unholy rows). The Labour Party had quietly dropped its belief that the state should directly involve itself in the productive economy.

The manufacturers that the Tories inherited in 1979 were a mixed lot. They included the giants nationalised by Labour as part of its industrial policy: British Steel, British Aerospace and British Shipbuilders (these last two had been state-owned for only two years), as well as Cambridge Instruments and Inmos, both half-hearted attempts to encourage the white-hot technological revolution. Other companies were lame ducks, sheltering under the wing of the National Enterprise Board. The biggest were British Leyland and Rolls-Royce. Finally, there was the oddball Royal Ordnance, manufacturer of armaments and long-time government employee.

Most of the government's charges were in poor shape in 1979. Only one company, the defence and electronics group Ferranti, had responded well to the NEB's treatment; it was quietly pushed back into the private sector. The other exception was British Aerospace, whose defence contracts kept it healthy. Just over half its shares were floated in 1981, and the rest in 1985.

The other companies were trickier. The government had a choice between selling them immediately in the hope that the buyer would be able to turn them round, and hanging on until they were healthy enough to be floated or at least make a decent price. With the smaller companies, it had no difficulty making up its mind: they were sold to whoever would take them. Fairey Engineering, Cambridge Instruments and Inmos were all offloaded.

That left the four big boys, Rolls-Royce, British Steel, British Leyland and British Shipbuilders, as well as Royal Ordnance. By floating Jaguar in 1984, the government gave notice that it was prepared to sell them off piecemeal if need be. The steel company and Rolls-Royce could not really be chopped up – their activities were homogeneous and in any case both seemed to be moving in the right direction. But a hard-headed decision was made to liquidate British Shipbuilders gradually through sell-offs; Graham Day was given a tight programme, and set to with ruthless efficiency. British Leyland was given more leeway – but not much. Even though its internal changes were clear enough to see, it was still making a loss and remained an irritant to the prime minister, to whom it was a symbol of past failure. She wanted it gone at any cost.

The stripping-down of British Leyland, now Rover Group, began in earnest a year after the Ford/General Motors fiasco: the politically safest option, sale to the management, was pursued wherever possible. First to go was Unipart, sold to employees in January 1987 and renamed UGC. In the same month, Leyland Bus was bought by its management; it was later sold to a foreign company, Volvo, without a murmur of political discontent. Similarly, Daf of Holland was allowed to buy 60 per cent of Leyland Trucks and Freight Rover in April; politicians could comfort themselves with the thought that the buyers were at least European. Eight other subsidiaries, including the software house Istel, Beans Foundries, Llanelli Radiators and a variety of foreign businesses, went during the year. All that was left was the volume car producer, Austin Rover, and Land Rover. Rover Group's workforce was now down to 42,700, compared with 73,000 in 1986.

In May 1987, Rolls-Royce was sold off to the public with the well-practised slickness of a big privatisation. For the government, it was a success. At the beginning of the year, City analysts thought the aero-engine company would float for £500 million; the eventual price, admittedly at the peak of the bull market, was £1.36 billion. It was almost ten times oversubscribed, with the majority of workers ignoring their unions' instruction to boycott the flotation. The company was doing very well indeed. Having lost £200 million in 1982 and 1983, it managed a £120 million profit in 1986, having captured 20 per cent of all turbofan orders from Western airlines. Ironically, its recovery was based on derivatives of the RB211 engine, the development of which had dragged it down in 1971. With a £3.2 billion order book, streamlined manufacturing operations and a reasonably solid position as one of the big three aero engine manufacturers, Rolls-Royce seemed to have a secure future.

In April 1987, Royal Ordnance was sold to British Aerospace for £190 million. Given BAe's position as a defence contractor, the fit seemed natural enough. But when, on March 1 the next year, it announced it also wanted to buy what was left of Rover Group, many eyebrows were raised. On the face of it, this seemed a strange deal, lacking industrial logic and turning BAe into even more of a ragbag than it already was; companies should, everyone knew, stick to their knitting. Furthermore, the link-up was announced in the same month that BAe announced its first ever loss, of £150 million (mainly due to the strong dollar), and Rover announced its first profit for five years. What was BAe's new chairman, Professor Roland Smith, up to?

The answer that convinced the initially horrified City was simple. Although British Aerospace had sales of more than £2 billion a year and, bar the odd currency disaster, healthy profits, it lacked a firm cushion of assets on which it could, if necessary, fall back. Rover, on the other hand, had a vast property portfolio. If Smith could get hold of Rover cheaply enough, he would be adding great muscle to BAe's balance sheet. When the City saw this, it started to applaud.

There were other advantages. BAe reckoned that Rover was now back on its feet. It predicted that profits would rise well above £100 million in the next three years, and possibly above £200 million. While Rover's five-year investment plan would absorb £1 billion, all but £200 million of that should be generated internally. Finally, there was some truth in the argument put forward by Day, and dismissed derisively by the City and press, that there was industrial sense in the merger. The point was that BAe's apparent might did not reflect the two groups' relative manufacturing capabilities: there was much that Rover could teach its new parent. "To spot the synergy," said Day, "you must focus on the manufacturing process, not the product."

BAe's strength came from its defence side. Unlike GEC or Plessey, it was not too sensitive to changes in the British government's purchasing policy because it was a highly successful exporter. But this constant fountain of cash had allowed it to get away with management practices that would have spelt death or takeover to many normal companies. Privatisation in 1981 – when the government had sold just over half of its holding – had made little difference, although approaches from both Thorn-EMI and GEC had given food for thought in 1984. It was only after the arrival of the ex-admiral Sir Raymond Lygo in 1985 that its deeprooted problems were tackled. He could see that BAe was overmanned, and that it was still riven by divisions between the old pre-nationalisation groups: its structure was not so much one of controlled autonomy as of a loose grouping of independent states. His first move was to destroy these old divisions by pulling all power into his own headquarters; at the same time, he stripped away layers of management, and started to bring consultants in to rationalise and more particularly to speed up production.

But in 1988 Lygo – now with the added incentive of red ink on the balance sheet – decided he had to cut costs by at least another third, and he needed all the help he could. There was much that Rover, which was a master of streamlining, might be able to offer.

Lord Young, the secretary of state for trade and industry, leapt at the proposed link-up. Although other groups, including Ford and Volkswagen, had expressed an interest, he gave exclusive negotiating rights to BAe. Purchase of Rover by another car company would, he knew, be likely to lead to embarrassing closures; and, of course, BAe was British. In terms of domestic politics, it was an excellent solution.

But he also knew that the stumbling block was not in Britain, but in Brussels. The European Commission's competition directorate could block a deal that was considered to give an unfair subsidy – the French government had already been forced to cut back its aid to Renault. To test the water, the first proposal was indeed generous: BAe would pay £150 million for Rover, and the government would inject £800 million, to write off Rover's debts and to provide extra for investment. The EC duly vetoed the deal, but by July Young thought he had a compromise, with the cash injection brought down to £540 million. Smith, who knew just what a strong position he was in, refused to sign the document and, an hour before the deal was due to be announced to parliament, Young had to throw himself into frantic negotiations with the wily professor. During these, he agreed to a variety of extra "sweeteners", including a delay in the date that BAe would have to pay the £150 million, and further direct cash subsidies. Had the EC learned of these, it would certainly have raised objections. But, by the time they became known, at the end of 1989, Rover was firmly integrated into BAe. The government had thrown the free market to the winds; in the process, it had produced a solution that had a good chance of industrial success.

As soon as the deal was signed, in August 1988, Lygo moved on to the next stage of his restructuring process by starting to decentralise BAe again. He created eight new divisions, one of which was Rover Group; he also announced tighter control of suppliers and a smaller headquarters. For all its strength, BAe had a long way to go before it would catch up with the best British manufacturers. It would be interesting to see whether Rover could help it.

Anyone searching for clues as to the likely success of the BAe/Rover merger would have done well to look at two companies that had recently come out of state ownership into the arms of private companies. One story involved Rover Group, the other British Aerospace. Both gave cause for optimism.

Leyland Trucks was taken over by Daf in April 1987. It was a rôle reversal of the sort that surprised no one any more: the upstart Dutch company, which started making lorries only in 1950, taking over the group that then dominated the European market. Daf was not yet a giant but, with 10 per cent of the European large truck market, it had a solid base.

The government, in its enthusiasm to push the deal through, agreed to write off £500 million and to allow the closure of the Scammell truck and Leyland engine plants. Despite this, there was a buzz of scepticism over the deal: analysts declared that the group would lose overall share and that rival makers would benefit from the erosion of Leyland's British image. Daf never had any such doubts: it saw the fusion of its own European sales network with Leyland's modern facilities and light truck range as a perfect fit. And so it proved.

Whatever pride the people at Leyland had lost in the takeover was amply compensated for by an extraordinary sense of liberation. The merger released a commodity that had been hard to find under British Leyland's umbrella of gloom: optimism. In the ten years to 1985, the workforce had fallen from 35,000 to 8,000. "All these guys were sitting there keeping their heads down, waiting for the axe to swing and catch them," says Martin Hayes, Leyland Daf's communications director. "To overnight become part of a dynamic growing organisation was a real fundamental change. It freed their horizons." From the day of the merger, Leyland Daf made money. All the hard work on the MRP2, quality and new products suddenly started to pay off. Over the next year, the Lancashire factory started to hum. Production jumped from 35 to 73 vehicles a day, productivity tripled. In 1986, Leyland sold 220 trucks in Europe; in 1987, it sold 3,000. Within two years, Daf had started to transfer production from Holland to Leyland.

Sometimes a government, however Tory, cannot be as ruthless as a private owner. After British Aerospace had bought Royal Ordnance, it made a decision that Mrs Thatcher could not have countenanced – to abandon the first small arms "manufactory" in Europe, the famous Royal Enfield works; 1,200 people would lose their jobs. This was BAe showing its new cost-conscious colours, and it was no coincidence that the man it hired to streamline the Ordnance production had previously been in charge of manufacturing planning at Rover Group. Peter Summerfield came out of the same mould as John Parnaby of Lucas – half academic, half businessman; a type much needed in manufacturing.

BAe's immediate problem arose from the losses that were piling up at Enfield. The factory was half way through the first part of a fixed price contract to supply a new rifle, the SA80, to the Ministry of Defence. Each gun was costing 30 per cent, or £100, more than it was being sold for; deliveries were eighteen months late. The new owners had two months to decide whether to take up the second part of the contract or to pull out: if they did the latter, they would in effect be getting out of the rifle business.

The Enfield factory suffered from an extreme form of traditional British ills but had, until the new rifle order, been able to hide its deficiencies. On winning the new order, which required unprecedented accuracy and complexity of manufacture, its management had decided that automation was the answer, and had brought in unfamiliar new equipment which in turn had led to a high reject rate. At the same time, the factory was breaking all the rules of efficient production: there were too many management layers, products and indirect employees, production of too many parts in-house, restrictive working practices and poor layout.

Along with its august history came entrenched traditions. Just as morale was high at Leyland Vehicles in the 1970s, it was holding up at Enfield. Uniquely among the state-owned manufacturers, Royal Ordnance had a form of monopoly: there had been no need to make the unpleasant changes that had so disturbed the rest of industry. BAe commissioned PA Consulting Group to do a study, which blew away all the complacency. It concluded that in theory all necessary changes could be done at Enfield but that they would be difficult to push through against established practice, and against strong union opposition. In August 1987, the company announced that it would go ahead with the rifle contract, but only by closing Enfield. By the following October, the weapons had been redesigned and were being produced at a profit in Nottingham: there had been no break in supply.

Even if production had stayed at Enfield, the factory would have been cut right back. That was because Summerfield and PA decided that they could transfer Japanese-style principles of car manufacture to guns. In particular, they decreed that instead of making 200 components, the new plant would make twenty – the high-value parts that could not be bought elsewhere. A small number of staff were moved up to spare space at RO's heavy gun factory in Nottingham. There, £15 million later, the antithesis of Enfield was created, with a flat management structure, flexible working practices, statistical process control, a zero defect policy, and a manufacturing design that meant that no expert skills were needed to assemble the gun properly. With overhead costs down by 50 per cent, output up by 45 per cent, productivity up three times and inventory down by 80 per cent, Royal Ordnance could do something it had never done before: export. It no longer had Enfield the place, but it had Enfield the brand name – and that, commercially, was much more useful.

It also had a political furore. When BAe paid £90 million for Royal Ordnance, many people said it had paid too much. Now, it stood to at least get its money back from selling the Enfield site. By developing it, as well as the nearby Waltham Abbey site, the company could, it was reckoned, make as much as £300 million. Was this, the Opposition wanted to know, deliberate asset stripping? The answer was no – but it did show, as also did the Leyland Trucks and Rover Group deals, that the government either had only the haziest notion of what had been going on inside its lame duckery, or preferred to turn a blind eye to the potential strengths of its inmates when it came to turning them out.

By the end of 1988, the fate of the other two members of the duckery had been decided. One lived, one died.

In the autumn of that year, a study found that British Steel was the lowest cost steelmaker in the world; when financing costs were taken into account, it even beat the Koreans. In December, it was floated for £2.5 billion. Ever since MacGregor had given his managers targets and told them to get on with meeting them, the corporation's productivity had been rising. In the five years after the great strike of 1980, three fifths of the workforce, 116,000 people, had lost their jobs. But at the same time, £1.3 billion had been poured into the five main plants. In 1987, 52,000 people produced as much steel as 166,000 had done in 1979: it took 6.2 man hours to produce a tonne of steel, compared with 13.2 in 1980. At the Port Talbot plant, which had received a £400 million refit, the figure was 3.7 – although the concept of man hours was a difficult one, because there were so few men. The great factory was almost totally automatic.

If BSC was set apart by its extraordinary level of investment, its other techniques were quite familiar to those who had read the right textbooks. It moved towards steels that were tailored to the needs of customers. It gave almost 25 per cent of its workers' pay in the form of bonuses closely linked to quality. It spent £20 million a year on training. And pay negotiations were now handled by one committee rather than twelve. In the year ending in March 1980, BSC had lost £1.8 billion, earning itself a place in the Guinness Book of Records as the biggest ever loss-maker anywhere in the world. In the year ending March 1988 it made £412 million profit, and the next year, £593 million; that made it Britain's fifteenth most profitable company. Even Ravenscraig seemed to have a secure future, and Bob Scholey, now chairman, was eyeing up possible targets on the Continent.

In 1978, British Shipbuilders had twenty-eight yards and 86,700 employees; ten years later, it had four yards and 6,000 employees. After the warship yards had gone, the full horror of the merchant business became clear: a loss of £25 million in 1984 grew to £148 million two years later. By including closure costs, the loss was technically greater than the turnover.

Just before Christmas 1988, North East Shipbuilders Ltd, the yard at Sunderland that had brought together Austin and Pickersgill and Sunderland Shipbuilders, was closed. Merchant ships had been made there for 600 years, and it had some of the most modern equipment in Europe. NESL died in the midst of an undignified squabble; a Danish company had ordered twenty-four ferries, but complained of poor quality; British Shipbuilders, rejecting the Danes' claim, cancelled the order. As the rest of the country savoured its health, Sunderland sank further into gloom. British shipbuilding was not dead, of course – it was being kept alive by VSEL (which looked healthy enough) and the other yards that had been sold off (most of which did not). But British Shipbuilders, created only in 1977, had come to the end of its short and unhappy life.

That wasn't quite the end of the great government sell-off – there was even one shipyard left. In Protestant East Belfast, where unemployment was three times the national average, Harland & Wolff and its neighbour, the aircraft company Short Brothers, provided a fifth of the local manufacturing jobs. In 1988, the shipyard lost £43 million and Short's lost £21 million, so privatisation was a matter of finding someone to take them at any price. It was a politically delicate task.

The government had ushered H & W into its lame duckery in 1975; since then it had given it almost £500 million and cut the workforce from 10,000 to 3,000. Northern Ireland, or rather the knowledge that government money flowed like water to anyone offering jobs there, had attracted its share of exotic failures, of whom John De Lorean was only the best known. So when Ravi Tikkoo, a West Indies based businessman who had briefly been in the news in the 1970s for declaring he would build the world's biggest supertankers, gave the shipyard an order for the world's biggest cruise ship, many eyebrows were raised. But Tikkoo had proved himself by building the tankers, so the government made him a counter-offer: have a shipyard to go with your ship. He declined, and ended up withdrawing from the deal, unhappy with the level of subsidy the government was offering. In September 1989, a team led by the management paid £15 million for the yard. The bulk of the money was put up by the Norwegian shipping magnate Fred Olsen, but more than 70 per cent of the workers bought shares. The VSEL formula was being repeated.

Short's had built biplanes for the Wright Brothers. The state had owned it since 1943, when it was most famous for its Sunderland flying boats, but it became a lame duck only in the early 1960s. That was when the RAF chose the American Hercules instead of Short's Belfast Freighter as its standard transport aircraft: the failure caused terrible damage to the Irish company's balance sheet. Although it produced a series of successful products, including a commuter aircraft, missile systems and a range of components for other planemakers, it never recovered its strength and, as a hotbed of militant protestantism, the government was chary of offering too much support. But now, in 1988, ministers decided it was time for Short's to go, and lined up two bidders: one was a consortium that included GEC and Fokker of Holland; the other was Bombardier of Montreal. In June 1989, Bombardier was chosen: it paid £30 million, and the government made up for its past neglect by giving a dowry of £730 million. Bombardier, previously better known for its trains (including the new ones on the New York subway) had been involved in aircraft only since 1986, when it bought Canadair. One of the newest plane makers in the world was buying one of the oldest, but no one was in a mood to feel too humbled.

In June 1989, Leyland Daf was floated on the Amsterdam and London stock exchanges. British Aerospace reduced its holding from 40 per cent to 16 through the sale, but still made a £90 million profit. Why, politicians demanded to know, had BAe been allowed to buy Rover for so little, and why had Daf been handed the truck division so cheaply? Both were certainly wonderful financial deals for the buyers, but that was now irrelevant. Daf had already proved it knew how to make the most of the lorry company; would BAe now be able to do the same with its car factories?

A hopeful sign came the next month, when Honda announced it was taking a 20 per cent share in the Rover Group. The two companies had first come together in 1978, when Michael Edwardes was desperately looking for a small car to fill a gap in his product range. He found the Honda Accord, renamed it the Triumph Acclaim, and it was soon established as the most reliable "British Leyland" car ever built. Then came the 200, little more than another rebadging, the 800, a genuine 50:50 effort, and finally the new 200, which would be launched in October.

At the end of 1988, Rover had stopped producing the Legend (the Honda equivalent of the 800) at Cowley, and there had been speculation that the Japanese were dissatisfied with the British quality. Certainly, there was an extraordinary difference between satisfaction ratings for the Rover and the Honda in the States: the Legend regularly came top of the polls, while the Rover just as regularly came bottom.

Given Rover's sophisticated production systems, the car's apparent lack of reliability was baffling, but Honda was clearly convinced that the situation was now under control. Not only was it taking a stake, it was also allowing Longbridge to produce 40,000 Concertos (its version of the new 200) a year. It would produce its own engines at a new factory in Swindon.

For British Aerospace, this was a satisfactory arrangement – it was guaranteed the support of mighty Honda, the most technically advanced of the Japanese carmakers. When the new 200 was finally presented at the Earls Court Motorfair in October, road testers loved it. It was, they said, solid, like a proper European car, definitely not like a tinny Japanese one. After years of dithering, Rover had finally decided to move upmarket, and away from direct competition with the European giants. Would this be the product that would save Rover, as the Golf had Volkswagen, as the 205 had Peugeot? Or would the 1990s be as unpleasant for the accident-prone car group as the 1970s and 1980s had been? To put it another way, would Rover once again become an acceptable brand – maybe even as good as Honda? To those who had watched the motor industry for many years, it seemed that the world really had turned upside down.

# 21

# *Slipping and sliding again*

At the end of 1987, John Butcher, the industry minister, said this of British industry: "What you are looking at is the early stages of what could become an economic miracle of the type achieved in the Federal Republic of Germany in the 1960s." The world started to agree: the following May, *Fortune* magazine made "Britain is back" a cover story. They were courting disaster: suddenly, in mid-summer 1988, gloom and doom came flooding back. The interesting thing about gloom and doom, though, was that it put industry under pressure for the first time. It had a chance to show off its new muscles – if it had any.

The gloom swept in on three major fronts: industrial relations, exchange rates and the sudden braking of the economy. On February 8, Ford had its biggest stoppage of the decade, a two-week strike that cost the company £20 million. This was a big test: had the bad old days of industrial conflict returned?

The dispute was partly about pay, but more about changes in working practices. Although Ford had been a leading proponent of the theory of Japanese practice, and had managed to convert its suppliers to Japanese-style policies, it had had a terrible time trying to implement them itself. Dogged by good profits – a function of its well-designed and marketed cars – it could never hold a credible stick over its workforce. It was only in 1985 that it finally managed to sweep away some demarcation lines, and reduce the number of job titles from 516 to fifty-two. That pushed productivity up by 60 per cent in two years. Some plants, notably the former Merseyside delinquent

Halewood, had improved dramatically. The threat of closure, combined with a series of trips to the identically-equipped Saarlouis plant, had convinced unions and management that they had better try to work together. But Dagenham Assembly, the most advanced car factory there was when it opened in 1931 but now cramped and bad tempered, had taken over as chief source of trouble. It took twice as long to build a Sierra as Ford's Belgian plant at Genk, its quality was inferior, and it could not operate just-in-time delivery. By 1993, Nissan would produce as many cars with 3,800 workers as Dagenham did in 1988 with 8,500.

The proposals that were now being so fiercely resisted were: to introduce systems to improve quality, including quality circles (they had previously been abandoned in 1981); to break down the distinction between skilled and unskilled workers; to create teams controlled by highly-trained, but not necessarily technically skilled supervisors; to harmonise blue and white collar employment conditions; and to be able to recruit part-time labour to meet production peaks. In other words, Ford wanted to copy Nissan.

The unions knew well that these proposals aimed to "Japanise" Ford, and that was what they resisted. They did not want quality circles, because they bypassed shop stewards' authority; they wanted to preserve the "craft" distinction between skilled and unskilled men; they were incensed that a supervisor, who could be semi-skilled, might be telling a skilled man what to do; and, most of all, they rejected the idea of hiring part-time workers. This last proposal was dropped even before the strike began.

The strike was called off after 67 per cent of the workers agreed to a package that slightly improved the pay offer, and meant that the other changes would be brought in by negotiation rather than imposition – small shifts, but important ones. Ironically, the stoppage had been particularly effective because Japanese-style just-in-time single sourcing was already operating at Ford's continental factories. The Sierra line at Genk ground to a halt within a week for lack of a relatively small number of British-made components. Had the strike lasted any longer, all European plants would soon have seized up. That emphasised what those who had studied Japanese systems knew quite well, that they required perfect industrial relations to operate efficiently. Ford managers immediately started to think about moving from single to dual sourcing: a great leap backwards.

The troubles at Ford were not as depressing as they had at first appeared. There were still elements of old-fashioned crustiness – macho managers, inflexible unions – but the ease with which the dispute was settled once the element of compulsion had been removed was encouraging. Had Ford's managers read their Japanese textbooks more carefully, they would not in any case have tried to impose the new practices. At the Toshiba plant in Plymouth, where the first no-strike agreement had been signed (and which had never even had to use the arbitration mechanism), the management held referenda on all proposed changes in working conditions. It was also significant that workers at every plant except Dagenham Assembly ignored union leaders' calls to reject the new package. At the modern engine factory at Bridgend, the vote swung from 80 per cent against an agreement to 87 per cent in favour of it. Dagenham was now on its own – the only factory where industrial relations were still stuck in the 1970s.

The next month, some unambiguous gloom emerged from Ford. It announced that it was abandoning plans to build a £40 million car electronics plant at Dundee, which would have created 1,000 jobs. The company managed to secure a single-union deal, but shop stewards at other Ford plants threatened to black it unless workers were paid the same as in the rest of the company. The last-minute pleas of the TUC's general secretary, Norman Willis, failed to change their minds. Ford said it would build the factory in Spain instead.

The government's economic policy had not been much of an issue since 1985, when unemployment had reached its peak. In comparison with the other four major industrialised countries, Britain no longer wore the dunce's cap. Gross national product had grown by 5.2 per cent in 1987, faster even than Japan; industrial production was up 3.8 per cent, behind Japan and America, but above Germany and France. The unemployment rate was no longer the highest of the five – it had slipped behind France. Inflation was higher than in France, Germany or Japan, but below America's: in the early months of 1988, it slid to 3.2 per cent. Interest rates were still above the others, but were falling. Only two indicators were slightly worrying: the appreciating pound, particularly against the dollar (up from $1.47 to $1.87 in the year to January 1988), and the trade gap, £1.5 billion in 1987.

Overall, the picture was rosy, and the chancellor felt he could pursue his ideological instincts in the March Budget by cutting the top rate of tax from 60 to 40 per cent. He continued to pull interest rates down: in May they reached the lowest level for ten years, 7.5 per cent. In that month, too, figures showed that average house prices had risen by 20 per cent in the past year: the economy was booming.

On June 2, Mr Lawson decided he had made a mistake: he reinstated the half point cut in interest rates and the pound tumbled as economists started to talk about the economy overheating. Statistics had just appeared showing that consumer spending in the first quarter of 1988 was 6 per cent up on the previous year. It was also becoming clear that inflation was now rising again and when the May trade deficit was announced – at £1.2 billion, it was almost as big as the whole of 1987's – the government started to panic. The budget had put £4 billion into already wealthy consumers' pockets at exactly the wrong time. As trade deficits yawned ever wider, the chancellor used the only tool he allowed himself – interest rates – in an attempt to slow consumer spending down. In October, the trade gap was £2.2 billion, interest rates had reached 13 per cent, and inflation was running at 6.4 per cent. The economy seemed to be going out of control.

It was not interest rates but currencies that brought the first gales of a new economic winter. Just as textiles and aerospace had benefited enormously from the collapse of sterling from 1983 to 1985, so now they were being beaten by its rise. The problem was not so much that sterling was rising as that the dollar was falling: civil airliners, all priced in dollars, had pulled British Aerospace into the red in 1987. Now Asian textile producers with currencies linked to the dollar were swooping down on the British market as their sterling prices fell. Though the British textile industry had spent £3 billion on new equipment since 1980, it was still not capable of fighting off the new tidal wave of imports. Coats Viyella's profits dipped sharply in the first half of 1988 and in the autumn, Courtaulds announced it was closing four Lancashire mills. That was just a start.

At the beginning of August, 1988, it was Jaguar's turn. Twelve months before, Sir John Egan had just been pipped by Sir John Harvey-Jones as "Britain's most impressive industrialist" in a MORI poll of directors. Now, as he announced profits down by a half in the first six months of the year, the City found it hard to forgive him. The high pound was the immediate problem, but analysts started to notice other weaknesses in its fallen favourite. Industrial relations had never been completely harmonious, and the cars were still suffering from quality problems: the two were, of course, linked. Despite Egan's enthusiastic adoption of Japanese-style techniques – including successful quality circles – he had found his attempts to change working practices resisted at every point by the unions. Only in May, there had been a week-long strike after a store manager allegedly poked a shop steward in the eye. Widespread share ownership among the workforce had done little to cure the problem: the theories of the textbook were not making it into practice.

There was a reason, though, which also set Jaguar apart from the rest of industry. Quite apart from its dependence on America, it was unusual in having decided that growth, rather than cost cutting, was the way to increase productivity: production had gone from 14,000 to 50,000 in five years, and capacity had been installed to allow for 80,000. A completely new car, the XJ40, had been launched in 1986 and had brought with it new quality problems. Sir John had given himself an extraordinarily ambitious task: to rationalise and grow at the same time. It was not surprising that he found himself faced with more than an average number of obstacles. He was paying the price of boldness – a rare commodity in the cautious world of the 1980s.

Despite the gloom, a poll by MORI for *Business* magazine at the end of 1988 showed that managers in large industrial companies were sunnily optimistic. Sixty-two per cent thought their companies would increase capacity in the following year, while only 2 per cent thought it would shrink. Just over half thought orders would increase, against 10 per cent who thought they would fall. Smaller companies, monitored by 3i, were sensing that higher interest rates would eventually lead to a downturn, but expected output and employment to rise in the immediate future.

The reason for this jolliness was that most companies were working flat out – lack of capacity was more of a problem than shortage of orders. Profitability was still rising fast (topping 10 per cent in 1988) and managers were confident that the chancellor, who had brought them such prosperity, would quickly suppress this outbreak of inflation and bring interest rates down again. Some did not want to hurry him: they had great piles of cash, and chortled happily whenever the rate rose. In the micro-economy, the land of companies, there was a whiff of complacency in the air.

This was in sharp contrast to the mood in the macro-economy, where economists and politicians lived. When Mr Lawson had allowed consumer spending to let rip, he assumed that industry would respond to the new challenge, and would either satisfy that demand or increase exports to balance a flood of imports. Neither happened. As industrial output failed miserably to match the growth in demand, the trade balance tipped crazily. It seemed to many people that the question of whether manufacturing was lean and fit or thin and emaciated had finally been resolved in favour of the latter.

But it was not as simple as that. Much of British industry had come close to expiry in the recession at the beginning of the decade. It had come out of intensive care, then out of hospital. By 1988, it was coming to the end of its convalescence: it felt fine, but it was still not strong. Then the chancellor asked it to run a marathon. That was silly – but no sillier than saying that it would never be able to.

Sir John Egan, with his dash for growth, was an exception among post-recessional British managers. The last thing most of them were thinking about was building up their British manufacturing base. They were cutting, consolidating, reorganising, and if they grew, it was invariably by acquisition. Everything they did was guided by caution. Like squirrels who had just made it through a harsh winter, they were still busy hoarding even though spring had arrived. As cash started to pour in, they put it to one side, strengthening their balance sheets and cutting their debt. At the end of 1987, debt was averaging a deeply conservative 10 per cent as a proportion of equity.

Investment in capital equipment had been the chief victim of this caution. In 1987, it was still running at a lower level than in 1979. After a spurt in 1984 and 1985, following the 1984 Budget announcement that tax relief on investment was to be cut, spending had stagnated. In 1987, the amount spent on machine tools was actually lower than the year before. Greenfield investment by British companies was practically non-existent: whenever there was an announcement of a new plant, it was invariably American or Japanese.

The figures did not tell the whole story, of course. First, British companies *were* investing like blazes – in other companies, often abroad. For Lucas, for example, the acquisition of a fuel system company in Detroit was as much an investment as the re-equipping of a factory in the Midlands. When share prices had been going loopy, there had been no need to pay in cash. Second, many companies were not spending on new equipment because they were busy trying to make the most of what they already had, juggling machines into cells, getting quality right; generally, trying to be Japanese. But that excuse took them only so far, for Japanese capital investment was running at twice the level – as a proportion of national income – as that in Britain. The Japanese were trying to grow; the British were not.

It was unfortunate that the jump in interest rates coincided with signs of a change in this mood, that companies were moving from the first stage, recovery, to the next, growth. In 1988, spending on capital equipment rose by 14.5 per cent: investment, adjusted for inflation, just squeezed past the 1979 figure. That in turn put more pressure on the trade figures: imports of machine tools grew by 21 per cent during the year. Much of the British machine tool industry had, unfortunately, failed to make it out of intensive care: in 1960, the UK had more than 9 per cent of the world machine tool market, now it had 3 per cent. Britain's trade balance paid the price: the 1988 total was an alarming £14.5 billion.

## Slipping and Sliding Again

Between September 1988 and March 1989, Courtaulds Textiles announced the closure of thirteen factories with the loss of 3,000 jobs, about a tenth of its British workforce. The textile industry, despite heavy investment, had a long way to go; the swing in exchange rates forced the closure of factories that might otherwise have been made competitive. But it also showed up weaknesses in the industry that could only be partially cured by investment. A report by the National Institute for Economic and Social Research said that the British women's wear industry was way behind the German one in almost every respect: much of the blame was put on lack of training. "The resource constraint in textiles is management," says Jeff Hewitt, group strategy director of Coats Viyella. "It's better, but it's not good".

In the spring of 1989, Coats Viyella made a £395 million bid for Tootal. David Alliance, now knighted, had his eyes on Tootal's Far Eastern thread factories, especially in China. The Office of Fair Trading decided to refer the bid to the Monopolies commission – perhaps not surprising as the combined group would control 70 per cent of the British thread market. Alliance's case rested on the belief that the group should be considered in an international context: he needed factories spread round the world to protect himself against currency fluctuations. For once, the MMC agreed: at the end of October, it said the bid could go ahead. But by then, conditions had become so tough that Alliance reduced the price he was prepared to pay, and Tootal kept its independence.

Another sector that followed its own business cycles was high technology. Since the gloomy days of 1985, the computer and semiconductor market had picked up. There were individual disaster stories – Rodime, the Scottish hard disk maker, had overstretched itself and had come close to collapse; in 1989, the American company Wang closed down its Stirling University factory as a result of worldwide problems. But there were also stories of happiness: Compaq, a rapidly rising star from Texas, established its only European factory at Erskine, near Glasgow, in 1987, and expanded it twice in the next two years. Acorn, along with other companies, recovered from past disasters by concentrating on computer software. It realised that this was where Britain still had a comparative advantage. If the education system discouraged discipline, it allowed full rein for the sort of thinking needed to create original software. Others acknowledged this strength: when the Japanese company Canon was considering the establishment of a unit to develop software, its thoughts turned naturally to Britain.

Racal had been moving along a similar path since 1981, when it had become jittery about its reliance on defence sales to the Middle East. It had sold several manufacturing companies, and had concentrated on data communications, where software and back-up were the main profit earners, as well as on its Vodaphone mobile telephone system. By 1988, less than half its workforce of 36,000 was involved in manufacturing, compared with 80 per cent in 1980.

Three other high tech giants had hurled themselves into a period of frenzied activity since the upsets of 1985. At the beginning of 1989, Thorn-EMI managed to escape from the sector completely by offloading Inmos. For Colin Southgate, it was the 169th time lucky – he had been turned down by every company he had approached in the previous three years, before finally striking it lucky with a joint venture owned by Thomson of France, and the Italian SGS. In exchange, Thorn-EMI took a 10 per cent stake in the Thomson-SGS, now the biggest chip company in Europe. So ended Thorn's five-year high tech fling: Inmos had cost it £300 million. This was fine for Thorn-EMI, and the City was delighted that it had shed such a risky venture, but those who saw Britain losing one of its few real world-beaters shook their heads. Sales of the transputer had lifted off, and 200 Japanese companies were already using it in their designs.

Southgate's biggest acquisition had been of an American television rental chain, Rent-A-Center, in 1987. Now, apart from its lighting division, Thorn-EMI had abandoned manufacturing. There were two ways to look at the sale of Inmos. On the one hand, it was a sign that safety-first City-led companies would find it hard to keep up any sort of technological lead: swapping transputers for TV rental was hardly adventurous. On the other, as Southgate said, Thorn would have had to concentrate vast resources to build Inmos up to the point where it could survive against the Americans and Japanese. At least with SGS-Thomson, which had microchip sales of more than $1 billion, it would survive. To those who took a European view of things, it was not such a bad solution.

In January 1989, Lord Weinstock renewed his bid for Plessey. Working with the German group Siemens, he bought just under 15 per cent of its shares. Weinstock had not had a good three years since his last bid was blocked: the cancellation of the Nimrod Airborne Early Warning system had dented GEC's technical credibility, while profits had stagnated. But they had been extraordinarily busy years. Those who had criticised GEC for running out of puff watched amazed as it concocted one deal after another.

Weinstock had encouraged his divisions to make smaller acquisitions – that was how Hotpoint came to acquire TI's Creda washing machine business. More adventurously, he had urged them to move into the unfamiliar territory of joint ventures. A telecommunications company, GPT, was set up with Plessey; GEC's medical equipment division was brought together with that of Philips of Holland; a joint company was set up with American firms to make commuter aircraft engines; and a group was formed with CGE-Alsthom of France. This last, a power generation company, would be the biggest Franco-British company, employing 85,000 people.

Joint ventures were a quick and obvious way to internationalise, but they had not had a happy history in British firms. Dunlop-Pirelli was just the best-known of a number of failures: Weinstock himself had had an unsuccessful fling with Hitachi in an effort to produce television sets jointly. A study into joint ventures by the Stanford Research Institute said that it was the Japanese, who tended to press for them, that benefited most, simply because Europeans did not understand how they should operate. Too often, the Japanese got what they wanted – usually access to the European market – while their partners got little in exchange.

Plessey, under constant threat from GEC, had also turned itself upside down, if in a more conventional way. Sir John Clark had brought in a young managing director, Stephen Walls, who had been applying the latest nostrums to Plessey, trying to make the group bigger in fewer areas, and more international. The company had bought Ferranti's semiconductor business, expanded in avionics by buying companies in America, and supped with the devil by forming GPT with GEC.

In this international atmosphere, it was not surprising that Weinstock chose to get together with Siemens to bid for Plessey. The Monopolies Commission had blocked GEC's bid because it looked at the market in purely national terms. Weinstock thought that an international bid might force it to take an international view, and Siemens' Karl-Heinz Kraske, who took the European single market very seriously, was happy to join in. The two men worked out which bits of Plessey each company would have, and where joint ventures would be formed.

Showing that it too could play the international card, Plessey struck back immediately by helping to form a group called Metsun to bid for and break up GEC. It was headed by Sir John Cuckney, chairman of Westland, and initially involved Thomson, STC and General Electric of the US, as well as Plessey. It was a disaster: news of it leaked out, and as one partner after another pulled out, the whole bid collapsed within twelve days. Most galling for Plessey was the defection of General Electric, which announced it was signing a series of joint ventures with Weinstock. Finally, in September 1989, GEC and Siemens overcame Plessey, paying £2 billion for it.

Throughout 1989, the economy slipped and slithered. Interest rates crept up further, to reach 15 per cent in October, then Nigel Lawson resigned as chancellor. Very slowly, industry started to lose its cool, although the business climate indicators – what managers thought about the economy as a whole – were consistently more pessimistic than the individual indicators, which showed what they thought about their own companies' prospects. Among small companies, those most vulnerable to interest rate rises, the climate indicator plunged from the summer of 1988, but it was not until a year after that they started to think that their own output would dip. Current indicators remained remarkably strong: both output and investment in the first half of the year were up on the 1988 equivalents.

Nevertheless, the 1989 trade deficit was heading towards £20 billion. More than ever, the manufacturing sector needed to grow.

Suddenly, in the midst of the gloom, squadrons of fairy godmothers flew in from abroad, bringing with them vast new manufacturing capacity. The deficit on cars and parts was £6.11 billion in 1988, by far the largest item in the total deficit. That was not surprising: the British had bought 2 million cars a year, but had made only 1 million. Then, all in a rush, foreign manufacturers decided that there was only one place to build new factories – and that was Britain. By the middle of the 1990s, economists calculated, the UK would be back in surplus on the motor trade.

The year had not started well, with an announcement from Ford that it was stopping Sierra production at Dagenham. But, the company said, it was investing £160 million in a new engine production line at the same site. It would be operated according to full Japanese principles, with just-in-time delivery and single sourcing for 96 per cent of components. Ford was not prepared to put too much trust in its workforce: this was to be one of the most automated plants in the world.

Then in the spring the foreign bandwagon started to roll. Other Japanese companies had been watching Nissan's experience in Washington with great interest. They took a long time making decisions, but when they did, they moved fast. Just before Christmas 1988, Toyota had started looking for a location for a £700 million factory. In April, it announced it had found one. It was at Burnaston, near Derby, an area without grants or a deep pool of unemployed labour. Toyota was attracted by the more traditional factors that pulled car manufacturers to the Midlands: a skilled workforce and the proximity of components suppliers. At the same time, it said that it would build a £140 million engine plant at Shotton in North Wales.

The next month, General Motors and its partly-owned associate, Isuzu of Japan, let it be known that they would invest £70 million to build a four-wheel-drive vehicle at Luton. And in July, Honda said that it would not only let Rover produce its cars at Cowley, but that it would expand its intended engine plant at Swindon to build 100,000 cars a year. Britain had swept the board in the competition to bring Japanese car makers to Europe, and no longer even had to use incentives to make sure it won. The three giants could be producing 750,000 cars a year between them by the middle of the 1990s.

When Japanese car companies had set up factories in the United States, they had brought component companies with them: by 1989, there were 300 and economists calculated that Japanese investment had done more harm than good to the local automotive industry. Whether that happened in Britain was up to British subcontractors. If they all behaved like Tallent Engineering, there would be few problems – but they did not. The early Japanese arrivals, including Nissan and Komatsu, had struggled hard to get the quality and service they wanted. They were starting to receive it, but only through a slow and continuous process of education, exhortation and straightforward pressure. But British companies would have to pull their socks up fast, for there were already signs that other subcontractors – not from Japan, but from Europe and America – were lining up to take the business. A flood of declarations poured in, all from companies keen to cluster round this new node of activity, and not too disturbed by the low labour costs.

In April, Bosch, the German giant, announced it was going to build a £100 million factory near Cardiff to make alternators; Montupet of France said it would build a foundry to make cylinder heads in Belfast, and Faure, also French, announced it intended to make car seats in Britain. There was even a little inward takeover activity: Valeo of France bought Delanair, the biggest British car heater company, and the shock absorber division of Armstrong Equipment (which had recently escaped the grasp of Wardle Storeys) was taken over by the American Monroe Auto Equipment.

All this inward investment was more than balanced by British takeovers abroad. It was an inevitable result of internationalisation that the companies that were helping the British balance of trade most were foreign, while the likes of Lucas and GKN were doing wonders for the German and American economies. But it was ironic that just after the only big British-owned car maker, Rover, had lost its independence, Britain was becoming a major car-producing nation again. In 1988, one third of all Peugeot 405s sold in France were made in Coventry.

The new structure of the industry had a complex effect on trade flows. To an extent, internationalisation had suppressed direct exports. GKN, for example, exported half as much to the USA in 1988 as it did in 1970, yet its American sales were 67 per cent higher. But there was a jump in exports of high value parts that were made only in Britain. Even the notoriously nationalistic German car manufacturers were starting to place contracts with foreign manufacturers: in 1989, Lucas won a contract to supply BMW with 400,000 fuel injectors a year. Nissan was already exporting to Europe, to the chagrin of the French; even if the overall car market slipped back, direct car exports from Britain could only grow.

This pattern was repeated in other industries, such as chemicals, pharmaceuticals and aerospace, where there had been a shift towards more specialist, international production. These were the areas where British industry was doing best. A report by Oxford Economic Forecasting in May 1989 predicted that UK growth in the early 1990s would be led by manufacturing, and particularly by exports from the automotive and aerospace sectors. This contrasted with two other reports, from the National Institute for Economic and Social Research, which were scathing about the design and quality of British kitchen furniture and women's clothing compared with that in West Germany. The signs were that the new pressures had shown up a division across British industry: on the one hand, the companies that were internationally competitive; on the other, those that were still struggling to sort out basic problems.

A surprising statistic came through in the middle of 1989. The trade on televisions and video recorders moved into surplus in the first quarter of 1989, after several years of deficits. Every volume producer in Britain was now foreign-owned.

The combination of falling unemployment and rising inflation was bound to lead to pressures on wages, and more tests of the "new" industrial relations. At the beginning of 1989, unemployment fell below 2 million, while inflation was rising back towards 8 per cent. Workers had been receiving steady 7 to 8 per cent pay rises since 1983; in 1988 they had crept above 9 per cent. The combination of higher inflation, lower unemployment and much higher mortgage repayments was not a good one to encourage wage restraint.

In the summer of 1989, there was a rash of disputes in the public sector: docks, trains, underground, local authorities and the BBC were all hit by strikes. But these arguments were not about pay, but about management attempts to change working practices. In manufacturing, such tension as there was, was linked to pay. It became clear that the main threat was that there would be a concerted campaign by the engineering unions to have the standard working week reduced from thirty-nine to thirty-five hours – in effect, a hidden pay claim. All previous awards had been paid for by improved productivity or had at least been linked to changes in working practices. Now, the unions wanted a straightforward, unfettered cut: they were celebrating the end of the recession and, most importantly, testing their strength. The leader was Bill Jordan of the AEU, a prominent new realist: his aim was to show the employers that they could not have everything their own way, and his left-wing critics that he was quite capable of prosecuting a successful dispute.

Jordan knew that the modern mortgage-burdened worker was much less inclined to strike than he had been ten years before. So he borrowed a tactic that had been developed by unions in Germany as they too fought for a thirty-five hour week. Workers in selected companies would be balloted and, if the ballot was successful, they would be called out on indefinite strike, with their pay made up by a national levy. Once the companies capitulated, there would be further ballots – until a domino effect started to bring hours down throughout the industry. It would be striking without tears – at least for the workers.

Since the 1987 stock market crash, there had been increased interest in leveraged deals on the Wall Street model. Partly this was because lower share prices had increased the use of cash in bids. Partly, it was because ever more sophisticated American financing techniques had found their way across the Atlantic. Foreign companies used them most – Elders IXL tried, and failed, to buy Scottish and Newcastle breweries; a New Zealand company, Goodman Fielder Wattie, also failed to buy Ranks Hovis McDougall. And in the spring of 1989 Gateway, the food retailer, was bought by a debt-financed consortium.

But it was in July of that year that Wall Street arrived with a vengeance, when a small group of investors, including Sir James Goldsmith, bid £13.5 billion for Britain's fifth biggest company, BAT Industries. Apart from the equity, which would come from twenty-six very rich men (Goldsmith was putting in £250 million himself), the bulk of the finance would come from junk bonds: BAT would then be stripped back to its core business of tobacco, and the money raised used to reduce the debt.

Apart from the usual arguments about such takeover targets being burdened with potentially inhibiting debt, this bid raised another question. Whatever Sir James had to say about BAT's failure to stick to its knitting, the rationale was purely financial. When conglomerates were in fashion in the City, their overall price/earnings ratio was higher than the average of each individual part, hence it made sense to keep them together. When they were out of fashion, as now, it made financial sense to break them up. Industrial or commercial logic had no place in the argument.

Nevertheless, Goldsmith did make one important point when attacking BAT's management, which had certainly not had the drive of the new-style conglomerates. "In my view," said Sir James, "conglomerates have failed in the long run. Their success comes in the lifetime of the people who have put them together." Interestingly, he is supported by one of the great conglomerators, Nigel Rudd of Williams. "If Brian McGowan and I disappeared," he says, "the basic business would run on for a number of years. But the dynamism – why don't we do this or that – would be gone. You would get perfectly good managers just managing but not being dynamic."

The point here is emphasised that modern conglomerates have more in common with the old-fashioned one-man companies – the Alfred Herberts, the Austins, the Morrises – than they do with other modern companies. Companies such as Lucas or ICI can be run like machines by professional managers; Alfred Herbert should have been, but it was not. But with the predators, as with all entrepreneurs, the vital spark had to come from one or two people.

In the autumn of 1989, Rudd and Williams gave another sign of dynamism – by announcing that they were starting to break Williams up. The car dealership companies were, they said, going to be floated off as a new company, called Pendragon. Sir Christopher Hogg at Courtaulds followed the same line: the textile division would become a separate company. It made sense for companies to break themselves up before someone else did. But what would happen if the process went as far as it had in America? With LBOs, there was no limit at all to the size of a bid. A few people, quite seriously, talked about the possibility of ICI being broken up. What would happen then to its seedcorn businesses, its hopes for the twenty-first century? The last vestiges of long-termism seemed to be under threat; it was alarming.

In late summer, the squeeze on spending started to bite hard. The earliest victims were retailers – particularly the clutch that had been subject to massive management buy-outs in the previous two years. High interest rates hit them twice – first as consumer spending slumped, and second as their own interest repayments rocketed. MFI, the biggest ever MBO, had to ask investors for more funds.

In August, Hoover made 200 people redundant because of the drop in consumer demand; London Brick, owned by Hanson, cut 600 jobs as house sales fell; Eatough, a privately-owned Leicestershire shoemaker, called in the receiver. The numbers were hardly comparable with the August nine years before, when 20,000 redundancies were announced in one week, but they did indicate an alarming reverse in corporate fortunes. For once, the South was hit harder than the North: the main mechanism for squeezing spending was through higher mortgages – and all the highest mortgages were in the South.

Business failures in the third quarter of 1989 were 23 per cent up on the year before; they were concentrated in building and construction, clothing and textiles, furniture and the retail sectors. Manufacturing output was growing at 3 per cent, less than half the 1988 rate. There was talk of a recession, with national income actually falling.

John Ashcroft of Coloroll was deeply gloomy. Consumer spending had, he said, dropped more sharply than it had in 1980. "We are stunned," he said. "Some sections of the economy are already in recession." As Ashcroft froze his investment plans, it became clear that his analytical skills had for once failed him. He had been caught out by the downturn, and at a time when Coloroll was still burdened with debt from the Crowther purchase. The City fell out of love with him, declaring that he spent too much time on extracurricular activities, especially the Prince's Trust, and not enough on running his business. It seemed that Ashcroft had caught an old English disease – Establishmentitis – at the grand old age of 40. Williams Holdings bought a 1 per cent stake in Coloroll: could the shake-out among the predators be about to begin?

In October came the first big collapse, of Hinari. It was a true creature of the 1980s, founded only in 1985 by an Englishman, Brian Palmer, and given a Japanese-sounding name to create the right image. The ploy worked: in 1988, sales rose by 75 per cent to £70 million, giving the company 10 per cent of the British 14-inch television market. Only in April, it had moved on from being a pure importer of electronic products to assembling television sets at Cumbernauld. Palmer put the blame for Hinari's downfall on the slump in spending.

At the other end of the industrial scale, Glynwed International announced it had just managed to beat its promised 20 per cent earnings growth target in the first half of 1989 – but it looked as though it would miss in the second half. That would be the first time Gareth Davies had missed his target since he announced one, in 1984.

Meanwhile, the pressure they had imposed had done spectacular things: Glynwed was now a very large engineering and distribution company, a touch larger, in sales terms, than Williams.

At the beginning of November, three factories owned by Rolls-Royce and British Aerospace were brought to a halt after the shop floor workers voted narrowly to strike for a thirty-five-hour week. But the two companies, fearful of losing their increasingly valuable skilled labour, did everything they could to avoid laying workers off at other plants. When, at the end of November, the Engineering Employers Federation then announced that it would no longer negotiate over hours on behalf of its 4,200 members, the union leaders realised the dispute would be anything but short and sharp. They would have to start negotiating firm by firm. Jordan made it clear that he would accept a thirty-seven hour week and on that basis, combined with a new productivity agreement, the workers at the Hillington Rolls-Royce plant voted to go back to work after six weeks. Some other companies had agreed to thirty-seven hours without strike action, but British Aerospace refused even to talk until the strike was called off. Jordan's attempt at militancy was running into the sand: the main effect of the strike so far had been to destroy the ninety-three-year-old national negotiating arrangement. That could only be to the unions' disadvantage.

The question of industrial relations had again raised itself, however, and there was a further ripple of anxiety as the Ford workers voted to take action in the New Year in support of a 10 per cent-plus pay claim. Just as industrial profits seemed likely to fall, the last thing managers wanted was pressure from wage costs, and Ford traditionally set the pace for pay agreements. Government ministers, terrified of a resurgence of wage-led inflation, made it quite clear that it expected the company to take a robust line.

But ministers also said that pay increases financed by productivity deals were quite acceptable. This left Ford in a difficult position, for the effect of the changes it wanted would be impossible to measure in the short run. Since the 1988 strike, some progress had been made on moving towards more flexible, Japanese-style, working practices. Now, the company wanted to take a few more steps, in particular by establishing cells, or teams, and by breaking down the demarcation between craftsmen and other workers. With more than £700 million profit forecast for 1989, it could clearly afford a much higher pay rise; what it could not afford was to lose momentum as it struggled along the Japanese path.

A question mark did hang over the companies that had supposedly solved their industrial relations problems by following this path, though. A late 1988 gathering of union officials from plants with single union agreements had expressed discontent that they were not being involved closely enough in discussions: once the deal had been signed, the shop stewards felt, some of the companies paid no more than lip service to mechanisms such as joint councils. That, if true, spelt trouble for the future.

Management buy-outs, the great success story of the 1980s, were coming under increasing pressure from high interest rates. Servis, a washing machine company that had been bought by its management in 1987, was paying the same in interest as it had been then, even though half the loan had been paid off. It was being hit both by its debt and by falling sales.

In Newark, Alan Bowkett of RHP had no such problems: he had swapped the main part of his management buy-in loan into a fixed 10.2 per cent rate in 1987. With the floating rate now at 15 per cent, he was saving £1.5 million a year. But that was a minor triumph compared with the deal he was currently concocting. In September, he had looked at the five-year prospects for RHP, and had decided they would not be served best by floating it on the stock market, as he had originally intended. City pressure for short-term profit would force him to starve the company of investment. So he flew to Japan, and persuaded the giant bearings group NSK to buy RHP for £145 million. Bowkett and his fellow managers would have £22 million to share out between them, they would still be in charge of an autonomous firm and its future would, they believed, be secure. It was an excellent deal for the company – and a deliberate snub for the City.

ICI's plunge into the red in the third quarter of 1980 had been one of the major shocks of 1980. The City was shocked by the group's third quarter results in 1989 too: profit had slipped to £306 million for the three months to September. It was still comfortably vast, but a sure sign that there would be slippages elsewhere. Brian Taylor announced that Wardle Storeys' profit for the year to August had more than halved: now, they were only about the average for industry. He declared that he would now concentrate less on margins and more on volume. As Wardle Storeys' share price slumped, Taylor found himself firmly in the City's doghouse: it did not forgive those whom it had worshipped, and had then proved themselves mortal after all.

In December, Constantine Folkes announced that he was selling Bar Bright to another private steel company for £4 million. Although the newly modernised factory had made a profit of more than £300,000 in 1988, Folkes explained that the high interest rates meant that it made sense to sell it to get rid of all the group's debt. Moves such as this were a logical but worrying concomitant of high rates – why should anyone invest in new equipment when they could make more by putting the money in a bank?

Despite these signs, the manufacturing base of the economy, which had been so badly damaged nine years before, was holding up reasonably well in the new climate. It was the sectors that had been untroubled by the 1980–81 recession that were feeling the pinch now. In November, figures appeared showing that receiverships had risen by 39 per cent in the first nine months of the year – they were now at the level they had been at the beginning of the 1980–81 recession. But the companies that were now collapsing were in building, retailing and motor dealing, and were concentrated in the South East. It would be interesting to see if these sectors responded to pressure as vigorously as the manufacturers had.

In the penultimate month of the decade, one of the great symbols of the 1980s renaissance lost its independence. On November 2, the Jaguar board accepted a £1.6 billion offer from Ford. Sir John Egan had gambled that he could make Jaguar big enough on the back of its American sales to allow him to spread out later. But the dollar had tumbled, and he had lost the gamble.

Towards the end of 1988, Egan had started talking to General Motors. GM wanted an upmarket brand name in Europe, and it agreed to fund Jaguar's new model development while contenting itself with a 15 per cent stake. It would have been a happy solution for those who wanted Jaguar to keep its independence. But on October 31, Nicholas Ridley, the virulently anti-interventionist new secretary of state for trade and industry, announced that he was relinquishing the Golden Share that allowed the government to block any takeover. Ford, which coveted Jaguar at least as much as GM did, pounced with a £1.6 billion bid. Egan preferred the GM option, but he could not turn down Ford's price – almost double the current share value. Within twenty-four hours, the deal had been done, and GM withdrew, upset.

Christmas shopping started late in 1989. Retail spending in October fell, when it would normally surge. "We are entering a demand recession," said a City economist. The toymakers sighed: they had seen it all ten years before.

# Part Seven

*Shaken but not stirred?*

# 22

# *Rebirth in Consett*

It is early summer in 1989 and, on a windswept moor some dozen miles from Durham, Consett is coming back to life. The town's reason for existence, the steelworks, has vanished. The great tract of land where it stood seems almost unnaturally empty, a little too carefully contoured to fit in with the surroundings. An oversized gateway and a small cluster of buildings – now home to the Derwentside Industrial Development Agency, DIDA – are the only vestiges of the mighty works. In an estate agency in the middle of the town, black and white "before and after" photographs show just how complete the transformation has been: in one photograph, a railway siding, electricity pylons, towering factory buildings; in the other, taken from the same spot, eight cows and a ploughed field. It is as nice a piece of reverse history as you are likely to find.

But what has been put in the place of the works, which covered the whole town in red dust, which could be seen from thirty miles away, and which employed so many people in Consett? The answer lies on the other side of the town from the old steel site, past rows of brown stone terraced houses and cut-price shops. Consett Number One Industrial Estate, despite the Stalinistic bareness of its name, is a hotbed of entrepreneurial activity, a monument to successful industrial regeneration.

The products made by the thirty or so companies there could hardly be further removed from heavy steel. They all fill niches in the best 1980s fashion. Best known, and getting bored with all the publicity, is Derwent Valley Foods, which makes the cleverly marketed and expensive Phileas Fogg snacks. Then there is Blue Ridge Care, making a new and superior form of nappy. Mark I'Anson's Integrated Micro Products, now turning over £4 million, and with a recently acquired subsidiary in Silicon Glen. And Duffus Clay, firing up elaborately decorated Victorian-style toilets and washbasins for the most up-market fitted bathrooms. Very little of what is made on Number One Estate will ever be bought by the people of Consett.

Despite the Dean of Wearmouth's dismissive coining of the word "consettisation", to mean the creation of low-paid, part-time jobs mainly for women, Consett's recovery has been real. The first new factories, including Derwent Valley Foods, did indeed employ women, and the men were scornful of what they called Mickey Mouse jobs. But that has changed. DIDA, the enterprise agency, provides the figures: by 1988, almost 3,961 jobs had been created, 71 per cent of them male and 95 per cent full-time. The unemployment rate had finally come down below the figure of the summer of 1980, just before the works closed. There are still more than 4,000 men out of work in Derwentside, but they include a lost generation of steelworkers who have never found another job. They work their allotments and will slip, eventually, out of the statistics.

Consett, which in 1981 had the highest unemployment rate in mainland Britain, has often been cited as a triumph of Thatcherite regeneration. But the free market had very little to do with its success. A better model would be the French post-war strategy, in which public and private organisations worked closely together to reconstruct industry rapidly. Nevertheless, Consett is not a bad model of the so-called "enterprise culture" at work, because the worker ants who have rebuilt it are typical of the new small businessmen of the 1980s: not natural entrepreneurs in the mould of Alan Sugar, and certainly not in the mould of Nigel Rudd or Lord Hanson, but ordinary business people who have been pressed, cajoled and tempted into taking a risk.

Three types of entrepreneur emerged in the 1980s. There were the people who started up a business on their own, and built it as best they could. There were the predators, who bought big businesses, and made them smaller. And there was Alan Sugar. Sugar did a disservice to the concept of the enterprise economy, because he made it easy for people to talk about "the likes of Alan Sugar". There were none: no one else started a company from scratch, and made it anything like as vast (Amstrad's sales in 1988 were £624 million) without buying other companies. He was a phenomenon but, like Halley's Comet, not one that we can expect to see very often.

The role of the predators will need to be studied further. That they were a product of the Thatcherite years is in no doubt. They could not have operated in a climate where earnings were taxed at seventeen shillings in the pound. Their single-minded pursuit of profit would not have been acceptable in the 1970s and, had the Tories not won the 1979 election, they would have taken themselves off to Australia, America, or wherever else they were allowed to do their dealmaking.

There is also no doubt that the predators were enterprising. But the real question is this: were they monsters that escaped from Thatcherism, only to wreak havoc on the economy, or were they important agents of change? Some people say that we should look at long-term research into past takeovers, which invariably shows that they have been harmful. One of the biggest, by Professor Julian Franks of the London Business School, analysed 3,000 companies involved in takeovers over thirty years, and concluded that the return to shareholders tended to drop. But this is a fruitless line of enquiry, simply because it is a nonsense to lump together takeovers in the 1980s with those in the 1960s. In the 1960s, the aim was to create size for its own sake; there was little rationalisation. In the 1980s, the aim was specifically to increase profit, with all the rationalisation that that implied.

In any case, the main effect of the predators, as constructive destroyers, is impossible to measure. One of the reasons for the American economy's dynamism is that there is great "churning" – companies rise and fall with far greater regularity than in Britain. There is no stigma to failure, and companies are allowed to collapse rather than struggle along in a semi-comatose state. Until the predators came along, aided by the newly-aggressive City, there was no real mechanism to clear out the chaff in Britain. Sometimes they took out good grain as well, but their role in speeding up the restructuring of industry during the 1980s must, on balance, have been a positive one.

But what of the third category of entrepreneur – the man or woman who starts up a company, and tries to make it grow? An economy can survive without predators; it cannot without a stream of new enterprise. Sir Keith Joseph bemoaned the lack of enterprise in Britain. Had it, as he intended it should, risen up during the Thatcher years?

By definition, entrepreneurial activity can be measured only by looking at large samples and broad trends. Many of the most valuable statistics come from the venture capital organisations, and particularly from 3i, the grand daddy of British venture capitalists.

One study for 3i[1] identified a number of factors that would point to a fundamental upsurge in entrepreneurial activity, or enterprise:

among them were a resurgence in new businesses, a decline in the share of large existing firms, and increased activity in the capital markets, especially in venture capital.

Domination of an economy by large firms tends to dampen enterprise by stifling an entrepreneurial environment. In the North-East, for example, the mastery of steel, coal, shipbuilding and a handful of large companies has led to a shortage of entrepreneurs. In the East End of London, by contrast (where Alan Sugar comes from), the market trader mentality has led to new firms popping up all over the place.

In the first twenty years after the war, the share in manufacturing output of the largest 100 companies in Britain rose sharply, from 25 to 45 per cent. It reached a plateau in the late 1960s and started to drop in the 1980s. The number of businesses with ten employees or fewer fell steadily from 100,000 in 1930 to a nadir of about 40,000 in 1965, then climbed sharply back to 100,000 by 1985. New company registrations paint a similar picture: a rapid climb from about 1968, with a dip only around the 1974 slump. The 1980–81 recession did nothing to slow the growth. In 1970, 30,262 new companies were registered; in 1986, the number was 114,831.

In the long run, then, it appears that there was a change in entrepreneurial attitude, not in the 1980s but in the mid-1960s. It was then that televisions became normal, that more people bought their own homes, that they started travelling abroad on holidays. In other words, there was an upsurge in individualism.

What the 1980s brought was a sea change in the environment for small companies. The government came along with a myriad of schemes; at first, most were aimed at high technology, later they became more general. Finally, in January 1988, the secretary of state for trade and industry, the former property developer Lord Young, announced that he was relaunching his department. Partly it was a marketing exercise. The media-conscious Young dubbed the DTI the "Department for Enterprise", and launched an elaborate advertising campaign to sell its services. But he was also redirecting the ministry, stripping away any service that might still support an industry in trouble, such as automatic regional grants, and concentrating all his fire on small and medium-sized companies. A £250 million budget was earmarked to finance consultancy studies for companies with fewer than 500 staff that wanted to improve their business skills. Local DTI offices were enlarged, and the civil servants in them told to become more like consultants.

Young's moves fitted perfectly with the philosophy laid down by

Sir Keith Joseph but that had, until now, been impossible to implement. The recession, the Westland affair and a series of other political ructions had seen to that.

Out of the recession came the enterprise agencies. Not corporatist exactly, but not based on the individual either, they were formed as a reaction to devastation by local communities. Councils, industry, bankers and other professional people came together to provide hand-holding and training service, similar to that given by BSC Industry, but with very limited funds. At St Helens, where one of the first agencies was established in the late 1970s, Pilkington – the great provider in that town – primed the agency with £50,000. Under the second Thatcher administration, the government, in the person of small industry minister David Trippier, started to encourage them itself. By the end of the decade, there were 300 agencies, and the London Enterprise Agency alone was advising or training 3,000 owner/managers a year.

The biggest change of all was in the availability of finance. Despite the anti-subsidy rhetoric, public sector finance flourished, especially in places like Consett. When Mark I'Anson and David Liddell were starting up in 1981, BSC Industry was an oasis of help in a hostile and cash-starved desert. In 1988, when Keith Stephenson was starting a company just along the estate from IMP, he found himself almost overwhelmed by financial support. His initial funding came from 3i but, with the help of a local accountant, he discovered an abundance of public funds. A regional development grant and regional selective assistance came from the DTI. Durham County Council provided two years' rate relief; the Common Market chipped in with a subsidy for each employee. BSC Industry came in with preference shares and an unsecured loan, and British Coal Enterprise (the mining equivalent) came up with another loan. Total funding came to just under £700,000, of which about £200,000 came from the public sector. "In retrospect," says Stephenson, "I would have gone to fewer people."

As the government made it clear that helping small industry was an absolute good (in the same way that helping exports had been, but was no more), even the clearing banks made at least a show of support. Greater professionalism was brought to bear; local managers were backed up with specialist units, and the small businessman became the hero of the bank advertisements. There was not much money in it, but it was good for the image.

Venture capitalism, by contrast, was good for the pocket as well, which was why so many banks decided to back it. In 1979, there were twenty venture capital organisations in the UK; in 1986, there were 126. The biggest, 3i, or Investors in Industry, was jointly owned by the Bank of England and seven commercial banks. Their growth was intimately linked with that of another phenomenon, the Unlisted Securities Market, which was founded in November 1980 by the London stock exchange. Venture capitalists operated by taking a share, perhaps 25 per cent, in a new company. Although they would be paid dividends as the company grew, they would always want to sell out at some point, so that the money could be invested elsewhere. Before the USM was founded, this was difficult: they might be able to sell their equity to another organisation, the company might be bought, or it might be floated on the full stock exchange. But the USM made the process much easier – companies could float on it with a shorter track record and without the expensive bureaucracy of a full float. As this was one of the easiest ways of becoming a millionaire, entrepreneurs required little encouragement to apply to join the market.

The USM grew up on the high tech boom that ended in 1985. By November 1984, 319 paper millionaires had been created by the 330 companies that had passed through it. Thirteen of the twenty most successful were in high technology. In 1985, the year of the high tech slump, the main market rose by 169 per cent on the back of the bull; the 100 biggest companies on the USM managed only a 17.5 per cent rise. But by then the story had got around that it was a good way to make a million, and even those who were sitting in comfy well-paid jobs started to jot down ideas on the back of envelopes.

Another way to get rich was through a management buy-out. These had grown in both number and size since Brian Taylor's £5 million deal for Bernard Wardle had pushed back the limits of financing in 1982. In that year, 170 MBOs were completed with an average value of £1.6 million; in 1986, there were 260 averaging £4.7 million; in 1987, MFI was bought out by its management for £718 million and in the next two years, there were half a dozen £100 million-plus MBOs. Even though the really big ones were concentrated in retailing, two out of three were of manufacturers. As the deals became bigger, so techniques became more sophisticated – first the management buy-in arrived then, in 1989, the great American leveraged buy-out.

By then, interest rates had dulled their glitter but, like the predators, MBOs had already fulfilled their role in the restructuring of industry. Early buy-outs were often the only way that a desperate company could raise cash – managers of subsidiaries were the only people prepared to buy them. Later, they became a mechanism by which companies could follow their 1980s textbook, stripping away extraneous activities, sticking to their knitting. Buy-outs were successful too: studies demonstrated that they consistently boosted productivity and profitability.

Venture capitalists provided the finance for MBOs but, by the mid-1980s, they were providing more than support – they were actively directing entrepreneurial activity. The management buy-in was their invention, it was a way of keeping up the momentum that MBOs had started. But it also showed the way the venture capitalists were thinking: that the future for British enterprise lay not in a pointless search for Alan Sugar clones, but in digging around in the middle ranks of large companies. They wanted to "liberate" managers who were able, but had gone as high as they could go, and persuade them that they should really be running their own affairs. "The common thread in success has been where really professional management is present," says Phil Goodwin, manager of 3i's Newcastle office. "We are trying to find the hidden entrepreneurs."

Alan Bowkett of RHP was a rather high-powered example of this new breed. Keith Stephenson, 42 in 1989 and with a passing resemblance to JR Ewing, was a more modest one. He came from Sunderland and, armed with a degree in economics from Lancaster University, had moved on to a standard big business career. He was on British Leyland's first graduate trainee scheme, in 1968 and 1969, and found himself specialising in the marketing of packaging products. That was where he stayed, and by 1984 he had worked his way up to become managing director of Sunderland-based Bowater Liquid Packaging. While there, he became convinced that there was a new product that was sadly underexploited – the plastic bottle for milk or fruit juice. If he was going to leave big business, he thought, he might as well do it now.

Stephenson had seen 3i's advertisements, and rang up the Newcastle office. "The manager told me to do some forecasts and put together a mini-business plan. I did some basic costing, and he said okay, now find yourself a good accountant. The accountant told me it would cost me £5,000 for a full business plan. I asked my wife what she thought, and she told me to get on with it.

"Building up the business plan was a question of burning the midnight oil. I showed how the market was changing from glass to plastic, identified specific customers, and got my pals at dairies to write letters supporting my idea. Then, the accountant put the whole thing together, and we sent out business plans to the various funding organisations asking for so much from each. Getting hold of grants was very simple. The DTI, bless them, were no problem: I was grilled for three or four hours, and then they said yes. BSC and the Coal Board went through the plan with a tooth comb. The redeemable preference share agreement with BSC Industry was *that* thick – a lawyer's paradise. Every time you blink, the lawyers charge you £500: I thought the professional fees would be £8,000 or £9,000 – they turned out at £25,000."

Nevertheless, with his tiny factory on Consett Number One, two green Italian extrusion machines, and eight workers (including an ex-miner and an ex-steelworker), Stephenson's company, PurePlas, started working in the autumn of 1988. A year later, it was running on four shifts, non-stop, and Stephenson had just ordered another machine. PurePlas was producing at the rate of 22 million bottles a year: it had 10 per cent of the British market. "I want to get 25," says Stephenson. "There's nothing like clout." Turnover was heading for £1.6 million, with £2.5 million scheduled for the following year. Consett had another success on its hands.

As 3i knows, Stephenson is statistically likely to survive – he is first a professional businessman, second an entrepreneur. Blue Ridge Care was founded by an American, David Langston. He had been sent to Britain by his company to set up a subsidiary: it pulled back, so he went ahead alone. At Derwent Valley Foods, Roger McKechnie came from the snack food business, and Ray McGhee from advertising. Duffus Clay was set up by Irwin Bamford, an accountant with an MBA, and Lyall Duffus, a ceramist. Professional managers dominate Number One estate.

By 1988, DIDA reckoned it had helped create 4,000 jobs in 200 businesses. Consett Number One is now just the flagship estate; there are thirty others in the area. BSC Industry had spent £1 million and helped 200 companies, although private capital had overtaken public money as the main source: £50 million came from venture capitalists, against £16 million from the public sector. As skills become scarce, DIDA has spent more and more time setting up or helping with training programmes. "Part of the problem with owner/managers is that they are the last people to realise that they need training," says DIDA's chief executive Eddie Hutchinson.

Consett is a long way from "critical mass", at which point growth is great enough to soak up the 4,600 people still unemployed. Hutchinson says: "We now have a new industrial base and, despite its strengths, it is still young. We want to get unemployment down to the level of Bournemouth." But if there is a recession, he notes with relish, it will not be Consett that suffers first: companies such as Derwent Valley

Foods and IMP have their own branch factories elsewhere.

Slowly, the cushions are being withdrawn from underneath Consett. BSC Industry has switched from unsecured loans to preference shares, the state-owned English Estates – which developed the factories – has stopped giving the two-year rent-free periods. Finally, Consett is rejoining the rest of the world.

23

# *Onward and ... upward?*

A couple of months before the decade ends, Bernard Robinson is feeling pleased, if a little exhausted. He has just been named North-East Businessman of the Year, and is now, thanks to a management buy-out, in control of his own destiny. Tallent Engineering is being flooded with work. Yes, he knows the British economy is tipping into recession, but he certainly is not. Puffing a cigar, he reels off some of his customers – Ford, Rover, Nissan, IBM, Goldstar, Samsung ... these last two are Korean firms that have recently set up microwave assembly operations locally. Tallent is turning over £30 million and has 540 employees, three times the number in 1981.

It is easy to see why Tallent is so popular with its clients. Robinson has taken all the latest techniques to heart, and loves them. At Christmas in 1986, he sent all his senior managers copies of a book on Japanese manufacturing strategy. Now, his factory is moving Nissanwards rapidly. He is obsessed with training, he wants to strengthen the role of supervisors, he is installing an advanced CAD system to link in with the manufacturers. Down on the shop floor, there is complete disorder – a sure sign that a cell system is operating. In one corner, sparks fly behind the wire mesh of a cage; inside, a fierce welding machine attacks the rear axle of a Ford Transit. In the middle of the factory, a series of meat hooks pass behind a long shield: small Nissan suspension parts hang from them, to be assaulted by robotic welders. The latest cell, not working yet, is where microwaves will be made for Goldstar – just-in-time production will be enforced, because there is room for only one microwave between each machine. It is a Japanese engineer's delight.

But the fact that Robinson is Businessman of the Year shows that he is out of the ordinary. Tallent's efficiency is matched by only a handful of large British companies, and by a spread of foreign-owned companies – American as well as Japanese. The principle of Japanese manufacturing has been universally accepted: every large manufacturer and many smaller ones know all about total quality, just-in-time, and so on. But there is still a chasm between theory and practice. "A lot of companies claim to have total quality management and just-in-time," says David Pearson, chief executive of the manufacturing consultancy and thinktank, the Strathclyde Institute. "But the benefits are not there. We perceive that the average company is still on the starting blocks. Virtually all the companies that are delivering just-in-time benefits are foreign-owned."

The problem is that for every one enthusiast, there are dozens who are half-hearted, and just-in-time cannot work properly unless it is fully accepted by every member of a supply chain: it has to become part of the industrial culture. Komatsu was deeply unhappy with the service it received from its British suppliers to start with, but was able to combine its educational skills with relentless urging to get what it wanted in the end. Smaller firms have had more of a problem. However much it has improved, British Steel still does not impress Bernard Robinson. "If your supplier is bigger than you, you cannot get JIT service," he complains: he buys steel from abroad, via a stockholder. Roy Ashwell, managing director of Mountfield, a Maidenhead-based lawnmower company, converted his factory to just-in-time in 1986. Extracting JIT delivery from his suppliers has been like pulling teeth. Large subcontractors always give priority to their largest customers, often at Mountfield's expense, while a raft of smaller ones have no intention of changing their ways. "They are just not used to sticking to delivery times," Ashwell says. "The majority of them are waiting for us to get fed up with this new idea and go back to bulk lorry deliveries." The best companies, he finds, are those run by young managers – or are foreign: he gets better service from Briggs & Stratton, his engine supplier in the USA, than he does from many British firms.

Manufacturers spent the 1980s trying to catch up with other people. It was a decade of unashamed copying, when the vaunted inventiveness of the British was put quietly to one side in a desperate effort to keep in the race at all. Cleverness was traded in for competence, because the people to beat were the competent Japanese. By 1990, the best British companies were travelling as fast as the best Japanese had been in 1980. Unfortunately, the Japanese would not stop accelerating. Just when they seemed to be going flat out, they had a nasty knack of finding another gear.

In 1981, Yamaha announced it was going to build a factory in Japan that would knock Honda off its perch as the world's leading maker of motorcycles. Honda sent up a cry: "*Yamaha wo tsubusu!*" – "We will crush, squash, slaughter Yamaha!" And that was what it did. Over the next eighteen months, it introduced or replaced 113 models, turning over its entire line-up twice. Yamaha, which had managed to produce a mere thirty-seven new models during the so-called H–Y War, retired with its tail between its legs.

A Japanese aluminium company was taking six hours to change its rollers every time it wanted to produce a different grade of foil. It started to train its workforce to cut that time: after five years, the changeover was taking nine minutes, and the target was eight and a quarter. A whistle would blow, scores of workers would descend on the rollers and go through their tasks, coordinated always by whistle blasts. "It was like synchronised swimming," says David Pearson, "but a bit faster."

In Canon's laser printer factory at Toride on the outskirts of Tokyo, much importance was given to the suggestion scheme. In November 1987, a Mr Itoh had his photograph on a board, because he had made 235 suggestions in the previous month alone – and every one had been taken up. At another Canon plant, where SureShot cameras were made, the managers aimed to get 500 suggestions per man per year – one every working morning, and one every afternoon. They were a little downhearted: right then they were getting only 443 a year. This was kaizen on speed.

In the two years following the Plaza Accord of September 1985, concerted government action forced the yen to double in value against the dollar: the aim was to reduce the Japanese trade surplus. There was a blip, as the profits of Japanese exporters plunged in 1986, then they started to recover. By 1988, the surplus was as alarming as ever. British moaning at an extra point on the interest rate started to seem a little feeble.

The ability of the Japanese to overcome a 100 per cent revaluation within four years did not show that they were supermen. They worked hard and long, and had a commitment to their companies that would be considered alarming in the West. No British worker would ever make more than 200 suggestions in a month – and few

## Onward and ... Upward?

would think it healthy if he did. But most of the time, there was nothing magic about the Japanese ability to find another gear – all they did was to sit down and think of the next logical step to keep ahead of the competition. Invariably, it was blindingly obvious in retrospect.

While the British were getting the hang of quality, the Japanese were forgetting all about it, or rather, they had come to take it, and the need for its constant improvement, for granted. They were concentrating on something else, the product. As Honda had discovered, a "variety attack" was at least as good a way to exhaust your opponents as a price war. But to make such an attack, you needed to sharpen up one familiar weapon, flexibility, and to introduce another – speed.

To base a strategy on product variety, a company had to be able to manufacture a wide range without losing economies of scale. That meant it had to have flexibility, to be able to change from one product to another as cheaply, that is as quickly, as possible. With mass production, the robot was the answer: it could treat each unit differently as it passed. But when products were made in batches, the only solution was to assault the set-up time, the length of time a machine was switched off between products. Until the 1970s, set-up times had been taken as given, even by the Japanese, but when Toyota decided it was going to switch to a variety-based strategy, its managers discovered that a combination of training – as at the aluminium factory – and automation could work wonders. In 1982, a study found that the set-up time for a heavy press was six hours in the USA, four hours in West Germany and twelve minutes at Toyota. This ability to switch models, combined with just-in-time production, meant that Toyota was able to offer any of 600,000 variations of its Corolla model with two-day delivery from order. For the first time in many years, car production was once again bespoke.

By 1989, Tallent Engineering was hitting thirteen minutes with its heavy press changeovers. Twin 800-tonne machines, massive blue arches, would be steadily punching out rocker covers for Ford Escorts in seven clean thumps. The machines would stop, the die, weighing 10 tonnes, would be rolled out and a new one, ready prepared, rolled in in its place. Straight away, the machines would start producing rear suspension arms for the Rover 200. Robinson was looking forward to getting his new machines – their dies could be changed in five minutes.

At the end of the 1980s, cutting set-up times had become a preoccupation among the better British engineering firms. It was a necessary part of a product-based strategy, but not a sufficient one. The missing element was the rapid design and development of products, and here the Japanese were in another league. They put their marketing people, designers, product and production engineers together in a huddle, gave them computer-aided design equipment, and told them to generate new products at ever shorter intervals. As the H–Y War showed, lead times could come crashing down: new models of cars, video recorders and CD players started to pour out of Japanese factories. Few Western companies even saw this as an area where they should be competing. There were exceptions: Xerox halved the development time of a new product from six years to three, while Rolls-Royce managed to cut the time needed to develop a new turbine blade from five years to two and a quarter. Other companies would have to do the same. Land Rover, which launched its Discovery in 1989, did well to develop it in three years – but it was a process that was started sixteen years after the last vehicle launch. Clearly, there would have to be a little more urgency in the future, or British companies would find their products were obsolete even before they were launched.

One of the great changes in the 1980s was that managers discovered whom they were pleasing. The answer, for publicly quoted companies anyway, was their shareholders. Profit was the spur. "I want to be remembered for increasing my earnings per share," says Brian Taylor of Wardle Storeys. Most managers who did well in the 1980s would agree with him.

That had the great advantage of simplicity: having only one goal, everything else fell into place – including, in principle, such things as pleasing the customer and even the workforce, without whom profit could not be made. But it had disadvantages too, which were well understood, and largely ignored.

In October 1985, just before the wave of great bids started to sweep in, David Walker, a director of the Bank of England, stirred up the "short-termism" debate. He pointed out that companies emerging from the recession needed to be able to make long-term plans to build themselves up and protect themselves against the next downturn. Fund managers, increasingly pressed to think in the short term, found themselves unable to turn down attractive offers from predators. Industrialists, reliant on keeping the City happy, seethed with frustration. "We can't say anything out loud," said one. "But we are all very worried by short-termism."

Walker's speech caused a brief stir, which soon died down. Nothing lasts very long in the City, and in any case there were powerful counter-arguments. First, it was demonstrable that the threat of takeovers did keep management on their toes – witness companies like John Waddington. Second, it seemed likely that many actual takeovers were beneficial, in contrast with those in the takeover boom of the 1960s. Then, mergers were mostly made for the sake of it, because big was beautiful. The characteristic of the 1980s merger boom was of efficient companies buying inefficient ones, and kicking their managements up the backside.

As British companies became stronger, and the mood turned against hostile takeovers, the short-termism argument became as unexciting as yesterday's toast. But the issues remained unresolved, and a trickle of companies started to take themselves away from the stock market and into the hands of friendly institutions. When Bernard Robinson led his buy-out at Tallent, he hunted around for institutions that would guarantee him long-term support: he has no plans to float the company.

This sort of thinking worries the more thoughtful denizens of the City. At a conference in July 1989, the director of institution funds for J Henry Schroder Wagg, Keith Niven, put forward a motion that "financial institutions should adopt a more enlightened approach to company performance and to the investments necessary to achieve this in the manufacturing sector". He pointed out that since the beginning of the 1980s, the average time that an institution held a stock had fallen from eight years to three and a half; Big Bang accelerated this short-termism by reducing dealing costs. "Fund managers are driven by the need to perform and show their trustees that their fund is in the upper quartile, and that their level of activity is high," he said. The trustees, he suggested, should stop looking so carefully at the quarterly performance tables – especially as they were quite likely to be businessmen themselves who in other circumstances would excoriate short-termism.

The most appealing argument in favour of the British system is that it is similar to that in America, the most dynamic industrial economy in the world. But Britain is not America: it does not have the flexibility or the attitude to thrive on constant corporate churning. Furthermore, the City of London is too big in relation to industry: too often the tail wags the dog. Interest and exchange rates are no longer determined by trade flows but by capital flows, with the result that companies' attempts to plan are liable to be tossed around like so much flotsam and jetsam. It cannot conceivably be good for the long-term health of the economy.

European managers based in Britain are astonished at the hostile environment British companies have to work in. The supportive stance of the German banks is well known, but the existence in Germany of supervisory boards with representatives of customers also makes sure that companies are encouraged to take a long-term view. They do not become complacent, because they have their eyes constantly on the competition; if they beat them, profits come naturally.

There is an excellent case for saying that the British financing system did a power of good to British industry in the 1980s. The great bull run, the rise of the predators, management buy-outs, management buy-ins, all helped to oil the wheels of the greatest industrial restructuring this century. There is still a need for more restructuring, but on nothing like the same scale. The question now is this: is the brutal British financing system the right way to ensure that industry starts to grow rapidly, as it should, or will it tend to restrict that growth? It is a question that is too easy to answer. The City system does not encourage investment for the long term.

What manufacturing needs is a German-type system of finance. If the plans for a single European market continue as planned, it should be able to get one. Could Deutsche Bank be the engine for British industrial growth in the 1990s?

And what of Europe, the single market, 1992? For companies that are not yet up to snuff, it will provide an extra dose of pressure. The history of the 1980s has been one of companies responding to pressure: first the recession, then the predators, still the City. Now, more open competition will provide another useful spur.

For companies that are already internationally competitive, the process will make little difference. ICI and Lucas cannot afford to draw a line around twelve countries and give them special treatment. They must always be seeking to beat the best in the world. In 1989, Sir Owen Green said this: "The story of the last thirty years has been of investment in the Asia/Pacific region – it is probably the most exciting thing that has happened to the world this century." Bernard Robinson makes the same point, from a different angle. "If you bother about Europe, you won't be around in five years time," he says. "You must be up with Korea or Singapore, places like that." Frightening though it may be, the Koreans call the Japanese the "lazy Asians".

It is difficult to say how wide and deep the changes in Britain's manufacturing base have been: it has yet to be put under real pressure. However, it is indisputable that manufacturing output grew faster in the 1980s than it did in the 1970s (but not in the 1960s). It is indisputable that it is no longer the great employer and is unlikely to be again. And it is indisputable that by most physical measures of productivity, British industry closed the gap on other Europeans.

There are still great weaknesses in Britain's manufacturing structure. It is much too small overall, and in some sectors it is still inefficient. These industries will need to follow the example of the handful of sectors that have done things right – those that have not only learnt how to make things efficiently, but that have also concentrated on products that are not too sensitive to price.

In the 1970s, too much of British industry was producing high-cost, low-quality, low-price goods. Now, in aerospace, automotive parts, chemicals, pharmaceuticals and a handful of other sectors, the combination has been flipped on its back: it is low-cost, high-quality, high-price. In other words, high-profit. The companies that have achieved this are those that will – backed up by massive research and development, and by judicious joint ventures (which will be a theme for the 1990s) – start to carve out a larger share of the world market. They will become increasingly international, spreading production and, to an extent, their development budgets, around the world. That is not a bad thing: a great commitment to research and development will be essential if they are not to be swamped by the Japanese, and they must use what resources they can.

Does this mean Britain will become an industrial outpost, full of factories belonging to other people and with no power over its own destiny? That is a danger. If foreign companies think of the United Kingdom simply as an alternative to Spain, a low-cost country with a robustly anti-union government, the possibility will become a reality. However, national self-denigration is a great British hobby – Canon has already set up a software development centre in Guildford, because it admires the flexible minds the higher education system produces. There is a place for cleverness in industry, as well as competence.

In the long run, there is a faint danger that that will be forgotten, and that cleverness will be ignored or suppressed at the expense of competence. At that point, Britain will become no more than a second-rate Japan. That day is a long way off. The key words in the 1980s were the likes of "professional", "structured", "logical" and "ordered", and in the 1990s they will have to be again. The British genius for improvisation and inspired short cuts – the art of the brilliant amateur – must, for the moment, be kept firmly in its box.

Meanwhile, as a glance at Japan makes clear, there is much to be done in the field of competence. But there remains a question for manufacturers: is Japan the right model? There is one alternative strategy – the technological one, beloved of and developed by the

Americans. Computer Integrated Manufacturing is, many companies in the USA have decided, the answer. At an Allen-Bradley motor control factory in Milwaukee, six people now run a plant that used to need 350. Each product chassis is given a bar code, which then "tells" the robots and other machines what they should be doing to build the product up: in effect, it is controlling its own assembly. Allen-Bradley claims that the factory is producing higher quality at lower costs than any of its rivals.

A handful of British companies have followed the CIM route. The factory where Rover makes part of its new K-Series engine can, for example, be operated with the lights out: there are no people there. But the technological solution is at best a partial one. Social and political implications restrict it to brand new factories, and there are sufficient disaster stories associated with high level automation to frighten most managers off. In any case, Britain still has relatively low wages – a good reason for sticking to the fundamentally attractive "Japanese" way of running a company.

The widespread acceptance of the Japanese gospel is based on the appeal of its basic premise: that the most under-exploited element in every company is its people. Give them more responsibility and more autonomy, and you will be rewarded with greater motivation and better results. Instead of 10 per cent of the staff thinking about ways of improving things, you have 100 per cent. The logic is irresistible.

There is a parallel to this in the management structure pioneered by Owen Green and Arnold Weinstock, and accepted as axiomatic in the 1980s. Responsibility has been pushed downwards, hierarchies flattened, subsidiaries given more autonomy. Again, you are making the most of people.

For the principle to work properly, though, people must both want responsibility – that is why good industrial relations are so important – and be capable of using it effectively. The skill of the predators has been to release hidden talent, to assume that there are good people trapped within their subsidiaries, and to seek them out. That technique, as BTR, Williams and Hillsdown have shown, is highly effective. But it will take a company only so far: to keep up the improvement, managers and shop floor workers must be trained. Training and education are the missing ingredients in British industry, and the new techniques of the 1980s have made them more essential than ever.

Here is an area where Britain is unambiguously worse off than its major rivals. The government produced a report in late 1989 showing that two thirds of employees had not received any training at all in the previous three years, and that such training as there was tended to be given as a response to short-term needs. Only 30 per cent of companies had any sort of training plan. The contrast with other countries was alarming: when the National Institute for Economic and Social Research looked at forty-five mainly engineering firms in Britain and Germany, it found that equipment was similar, but that productivity was between 10 and 130 per cent higher in the German factories. The reason, the Institute concluded, was that two thirds of the German workers were qualified craftsmen, against fewer than a third of the British. It also commented that German supervisors had much higher status, and were much better trained. In another comparison, NIESR found that France was producing two and a half to three times the number of technicians in mechanical and electrical engineering that Britain was. In offices, the situation was no better. Despite a desperate shortage of computer-trained people, a survey in 1988 showed that of the country's 1,000 biggest computer users, 400 provided no internal training at all. The story was repeated again and again.

After cutting back in the recession, British companies slowly started to increase their training programmes. In 1987, they were providing 15 per cent more days than they were in 1984 and, as demographic changes promised to squeeze the number of young people, training became an obligatory worry in the boardroom. The government's schemes were concentrated on young people and the unemployed, although in 1989, it followed its ideological nose by creating Training and Enterprise Councils. TECs would bring together businessmen, chambers of commerce and enterprise agencies to design schemes most suitable for local conditions. Industry liked the idea – its results would be tested in the 1990s.

Managers who followed the Japanese way invariably gave great emphasis to training; they noticed two things that set Japanese companies apart. First, it was always carried out in-house: managers would train their own staff, constantly, and repeatedly. Second, the Japanese did not believe in specialisation. Craft training, in the style of the old apprenticeships, was becoming ever less appropriate as fast-changing technology and cell-working made carefully nurtured skills obsolete. In general, British companies did not adopt in-house training, but they did think about ways of creating the new flexible worker. In 1988, Perkins Engines started on a £1.2 million programme designed to produce a 140-strong maintenance team by 1992: each member would be "multiskilled", or trained to do any shop floor repair task. Other companies took advantage of the service offered by the "new realist" unions. Both the AEU and EEPTU had established training centres, which were increasingly used to generate workers with a variety of skills.

General education was now more important than ever, and again the British were missing out. In Japan 95 per cent of children stayed at school until they were 18, compared with 32 per cent in Britain; A-level specialisation at 16 failed to provide the broad education industry wanted. Here, the government did make a number of changes. The most fundamental was the introduction of the core curriculum, in 1988: it was designed to guarantee a basic level of knowledge for everyone. Two years earlier, City Technology Colleges had been announced with great fanfare and a promise that twenty would be open by 1990. These schools, which would be partly financed by companies (Hanson, Lucas and BAT all signed up), would be designed to produce highly motivated youngsters with the skills and knowledge that industry needed. They would be worked harder, more in the manner of strict grammar or private school. Soon the scheme slipped behind schedule, dogged by opposition from those who saw the CTCs as élitist; by 1990 only three had been opened.

The idea of companies providing their own education programmes was well established among the enlightened Americans, such as IBM, but little used by the British. In 1989, both Ford and Lucas announced ambitious schemes. Ford's, which was to be run jointly with the unions, offered evening classes in subjects not related to work, as well as grants for employees who wanted to take Open University courses. The scheme at Lucas was similar: it offered language tuition, study for GCSE exams, and even post-graduate business administration and engineering courses.

Management training was weak – according to a 1986 report, more than half of British companies made no formal provision for it. But there was a more fundamental problem, indeed *the* fundamental problem of British industry. The anti-industrialism of the British had, it appeared, scarcely been scratched during this decade of frenetic industrial activity, and that was starving manufacturing of the talent it so desperately needed. The City had become glamorous, some of the dealmakers – the predators – had become almost famous, and young people had shifted politically to the right. Every television channel had established its own business programme, new magazines had started to pack the newsagents' shelves, and share owners had come to outnumber trade unionists.

Somehow, though, most of this had passed manufacturing by. A poll among students in 1989 found that banking and finance was the third most popular potential career. The media came top, education second – and manufacturing last. Only 7 per cent of the 1,000 students asked said they would even consider a career in it.

Attempts were made to raise the status of industry. Business in the Community, an inner city charity, organised visits by senior executives to schools throughout the country. ICI employed two former teachers to encourage as many children as possible to become scientists and think about industry. And Bernard Robinson took to visiting local comprehensive schools to tell children about the thrills of industrial life.

Demographic changes were ensuring that the task of attracting the right people would only get tougher. The number of graduates would stagnate during the 1990s, while demand for them would rise by 30 per cent. At the beginning of the 1980s, even bright graduates were struggling to find jobs. Already by 1989, they could pick and choose. With some banks offering £25,000 starting salaries, manufacturers were left wondering how on earth they could beat the "demographic time bomb". By the end of the decade, it had yet to explode – but unless something dramatic happened to change the anti-industrial culture, they would feel the full blast of it.

On Christmas Eve 1989, Simon Jenkins, writing in the *Sunday Times*, poured scorn on the people for whom the 1980s had been a golden age, those who "made money but did not earn it" through property, merchant banking, consultancy. "Who should have been the Man of the Eighties?" he asked. "I can tell you. He is busting his gut 12 hours a day to keep all the above in cufflinks. He runs a factory, manages a company, produces, serves, hires, fires, meets a payroll, answers for his bottom line. He carries on his shoulders a responsibility unknown to those outside business: to keep a team of individuals committed to production over time and against competition. One false move and he cannot take refuge behind ministerial responsibility or a professional smokescreen. He loses other people's jobs and money, and his own.

"Raising the esteem of management, the poor bloody infantry of the British economy, is simply the toughest task Britain has to crack in the 1990s," Jenkins wrote. He was right. Someone, somewhere, has to convince Britain's youth that there are worse places to spend a life than in the iron universe of industry. Manufacturers should remove their noses from the grindstone for a moment, and start explaining to everyone how clever they have been. They do not have a bad story to tell.

# Notes

**Chapter 1**
1 Martin Wiener, *English Culture and the Decline of the Industrial Spirit, 1850–1980*, Penguin, 1985.
2 Ibid., p. 18.
3 Ibid., p. 23.
4 Jonathan Wood, *Wheels of Misfortune*, Sidgwick & Jackson, 1988, p. 89.
5 Ibid., p. 94.
6 Wiener, op. cit., p. 144.
7 Ibid., p. 202.
8 Huw Beynon, *Working for Ford*, Penguin, 1983.
9 Wiener, op. cit. p. 109.
10 Ibid., p. 110.
11 Corelli Barnett, *The Audit of War*, Macmillan, 1986, p. 59.
12 Quoted in Wood, op. cit., p. 161.
13 Beynon, op. cit., p. 107.
14 Wood, op. cit., p. 112.
15 Graham Bannock & Partners, *Britain in the 1980s: Enterprise Reborn?* 3i, 1987, p. 17.

**Chapter 2**
1 Wiener, op. cit., p. 162.
2 Wood, op. cit., p. 220.
3 Boston Consulting Group, *Strategy Alternatives for the British Motorcycle Industry*, Cmnd 532, HMSO, 1975.
4 Wiener, op. cit., p. 201.

**Chapter 3**
1 *The Annual Register*, 1979, p. 26.
2 Ibid.

# NOTES

**Chapter 4**
1 Ian MacGregor, *The Enemies Within*, Collins, 1986, p. 99.
2 *Sunday Times*, 2.3.80.
3 Sir Michael Edwardes, *Back from the Brink*, Collins, 1983, p. 127.

**Chapter 5**
1 Dart Spring story from *Sunday Times*, 29.6.80.
2 *Financial Times*, 9.7.80.
3 *Sunday Times*, 31.8.80.

**Chapter 6**
1 *Sunday Times*, 1.1.82.
2 MacGregor, op. cit., p. 97.
3 Nicholas Faith, "BSC casts off its past", *Business*, July 1988, p. 48.
4 *Financial Times*, 15.9.81.

**Chapter 7**
1 MacGregor, op. cit., p. 102.

**Chapter 8**
1 Writing in the *London Standard*, 24.2.86.

**Chapter 10**
1 Anthony Sampson, *The Changing Anatomy of Britain*, Hodder & Stoughton, 1982, p. 354.

**Chapter 11**
1 Wood, op. cit., p. 241.

**Chapter 13**
1 House of Lords, *Report from the Select Committee on Overseas Trade*, HLP238, HMSO, July 1985.

**Chapter 15**
1 *Fortune*, 16.4.84, p. 31.

**Chapter 16**
1 Ingersoll Engineers, "It's not so much what the Japanese do ... it's what we don't do", *Ingersoll Engineers*, Rugby, 1982.

**Chapter 18**
1 Satashi Komata, *In the Passing Lane*, George Allen & Unwin, 1983.

**Chapter 22**
1 Graham Bannock & Partners, op. cit.

# Bibliography

Corelli Barnett, *The Audit of War*, Macmillan, 1986.
Philip Bassett, *Strike-free: New Industrial Relations in Britain*, Macmillan, 1986.
Huw Beynon, *Working for Ford*, 2nd Edition, Pelican Books, 1984.
Michael Edwardes, *Back from the Brink*, William Collins, 1983.
David Halberstam, *The Reckoning*, Bloomsbury, 1987.
William Kay, *Tycoons*, Judy Piatkus (Publishers), 1985.
Andrew Lorenz, *A Fighting Chance*, Hutchinson Business Books, 1989.
Ian MacGregor, *The Enemies Within*, William Collins, 1986.
Geoffrey Maynard, *The Economy under Mrs Thatcher*, Basil Blackwell, 1988.
Nick Oliver and Barry Wilkinson, *The Japanization of British Industry*, Basil Blackwell, 1988.
Thomas J Peters and Robert H Waterman Jr, *In Search of Excellence*, Harper & Row, New York, 1982.
Sidney Pollard, *The Development of the British Economy*, 3rd Edition, Edward Arnold, 1983.
Paul Usher, *Putting Something Back*, The Planning Exchange, 1989.
Peter Wickens, *The Road to Nissan*, Macmillan, 1987.
Martin J Wiener, *English Culture and the Decline of the Industrial Spirit, 1850–1980*, Penguin, 1985.
Jonathan Wood, *Wheels of Misfortune*, Sidgwick & Jackson, 1988.
Hugo Young, *One of Us*, Macmillan, 1989.

# Index

AEG 81
AEI 13, 81, 152
AEU *see* Amalgamated Engineering Union
AIMS *see* Advanced integrated manufacturing system
APV 238
AUEW 34, 41, 45
Abdullah, Raschid and Osman 108, 114, 243, 244, 251
Abell, David 95, 107–8, 114, 251
Accountants, greater role 17
Acorn Computers 82, 143, 275
Acrow 120–1
Advanced gas-cooled reactors 12
Advanced integrated manufacturing system (AIMS) 191–2, 198
Advertising industry 17
Aerospace industry 10, 13, 65, 134–5, 272, 280; and Government 20, 24; US 10; *see also* British Aerospace
Airfix 35–6, 71
Aldington, Lord 154
Alliance, Sir David 206, 207, 275
Allied-Lyons 202, 211, 214–15
Allport, Denis 61
Amalgamated Engineering Union (AEU) 228, 281, 308
America *see* United States
Amstrad 256, 290
Anderson, Eugene 127
Ardeer 44
Argyll Foods 203, 207, 209
Ariston 83

Armco 243
Armstrong Equipment 252, 279
Ashcroft, John 17, 20, 49, 115–17, 164, 165, 167–8, 245, 283
Ashley, Laura 135
Ashwell, Roy 299
Asset stripping 14, 95
Aston Martin 51, 117–18
Aston Martin Tickford 118, 214, 253
Atkinson, Sir Robert 88
Attlee, Clement 9
Austin, Sir Herbert 5
Austin Morris 25
Austin Motor Co. 13–14
Austin Rover Group (ARG) 139–40, 187–9, 209, 226; labour relations 137, 229–30; privatised 260–2
Automated guided vehicles (AGV) 192
Automation 82, 84; motor industry 187–92, 301
Aveling Barford 211
Avery 152
Aviation industry *see* Aerospace industry
Avon Rubber 242

BAT Industries 202, 281–2, 308
BBA 114–15
BET 246–7
BICC 121
BMW 107, 280
BOC scheme 232
BP 28–9, 244
BSC Industry 73, 151, 293, 296, 297

313

# INDEX

BSR 49
BTR (British Tyre and Rubber) 17, 90–1, 95, 97–9, 102, 103, 104, 164, 170, 172, 182, 251; offer for Dunlop 128–30, 215–17
BTR Nylex 217
Babcock and Wilcox 57
Babcock International 238
Baker, Kenneth 144
Baker Perkins 238
Balance of payments deficit 43
Balance of trade, report 154–7
Baldwin, Stanley 7
Ballbearing industry 13
Bamford, Irwin 296
Bank of England 80, 85, 127, 128
Banks, and small businesses 72, 293
Banro 214
BarBright 243, 285–6
Barber, Anthony 20, 26, 165
Barclays Bank 80, 105, 107, 127
Barker, Trevor 244
Barker and Dobson 245
Barnett, Correlli 8
Barr, Andy 187–8, 189, 226
Bathgate 137
Battenberg, J. T. III 209
Beans Foundries 260
Beatrice Foods 175–6
Beaver Group 51, 117
Beckett, Sir Terence 59, 158
Bedford Trucks 209
Bell's 201, 202
Benn, Tony 24
Bentley, John 20
Bentley Cars 179–80
Benz, Carl 5
Berec 99, 204
Berger 237
Berger, Jenson and Nicholson 250
Beveridge Report 9
"Big bang" 200, 215
Binns, Peter (Binns Cornwall) 212
Biotechnology 145, 174
Black Country, and recession 48–9
Blakely (ICI) 84, 175
Bloom, John 83
Blue Arrow 244
Blue Ridge Care 290, 296
Boeing AWACS 214
Boesky, Ivan 129, 202, 215
Bombardier 267
Bosch 279
Boston Consulting Group (BCG) 27, 28, 198
Boucher, Nick 119
Boulton and Paul 246, 247
Bowater 53

Bowkett, Alan 246–9, 285, 295
Branson, Richard 163
Bray, John 116
Bridon 57
Bristow, Alan 208
British Aerospace 34, 74, 135, 147, 207, 238, 259, 260–1, 272; flotation 139; Rover acquisition 260–2, 267; strike at 284
British Aluminium 243
British Coal 87
British Coal Enterprise 293
British Empire as market 9, 10
British Gas 145
British Leyland 24–5, 32, 34, 40, 44–5, 49, 59–60, 69, 84–5, 89, 137, 139–41, 169, 187–90, 200, 226, 258, 259–60; and Honda 76; labour relations 23, 41–2; and Land Rover 209–11; troubles 52, 55, 68, 78, 89–90, 138
British Motor Corporation (BMC) 13–14
British Printing and Communications Corporation 111
British Rail 145
British Shipbuilders (BS) 33, 44, 88, 139, 141, 258, 259; privatisation 265–6
British Steel Corporation (BSC) 12, 15, 33–5, 43, 44, 47, 60, 61, 70–1, 80, 87, 88–9, 133, 136, 138, 139, 162, 187, 197, 238, 259, 299; privatisation 265; strike at 39–41; troubles 52, 55, 56
British Technology Group 74, 76, 151
British Telecom 145, 258
Britoil 244–5
Brittan, Sir Leon 203, 208
Brown Boveri Kent 33
Bundy 243
Burmah Oil 181
Burnham, Chris 16
Business in the Community 309
*Business* magazine 273
Business Start-up Scheme 72
Businesses, size of 292
Butcher, John 269
Buxted 181
Buyouts (MBOs) 85, 284, 294–5

C5 electric tricycle 148, 150
CAD/CAM 188–9
CAV 22–3
CGE Alsthom 277
CH Industries (CHI) 117–18, 214, 252–4
Cable TV 144
Cadbury's 49
Cahill, John 97, 216
Callaghan, James, Lord Callaghan of Cardiff 21, 22, 28, 30
Cambridge Instruments 32, 138, 259

314

# INDEX

Cammell Laird 141, 258
Canon 87, 183, 275, 300, 305
Capital investment, low 239
Capitalism, fear of, in UK 7
Capper Neill 121
Carrington Viyella 206
Cars 165; sales of 46, 54, 90, 140, 278; *see also* Motor industry
Cartels 7
Castle, Barbara 16
Cell working 195, 229
Celltech 145
Centre for Policy Studies 31
Challen, David 113, 216–17, 254–5
Chamberlain Phipps 252
Chambers, Sir Paul 171
Champney's 202
Channon, Paul 208, 210, 211, 258
Chemical industry 7, 10, 23, 56, 58, 238, 280
Chesterton, G. K. vii
Citicorp Venture Capital 85, 247
City, The x; *see also* "Big bang"; Stock market
City technology colleges 308
Civil aviation industry 134–5, 238
Clark, Sir Allen 153
Clark, Graham 250
Clark, Sir John 153, 158, 277
Clark, Michael 153
Clark's Shoes 54
Clarke, Kenneth 255
Clegg Commission 43
Coates, Ken 115
Coats Patons 204, 205–6
Coats Viyella 207, 238, 272, 275
Coles Cranes 120–1
Coloroll 49, 108, 115–17, 167–8, 245, 283
Colston, Sir Charles 83
Company size 292
Compaq 275
Competitive achievement plans (CAP) 177–8
Computer boards 85–6
Computer industry 13, 53, 142–5, 148–52, 275–6; Germany 53; *see also* ICL
Computer-integrated-manufacture 189, 306
Computers *see also* CAD/CAM
Concorde 12, 135
Confederation of British Industries (CBI) 12–13, 42, 54, 59, 80, 134, 157
Conglomerates 96–7; *see also* BTR; Hanson Trust
Conservative Party, economic policy 1974 31; election victory 1979 3
Consett 35, 40, 47, 62, 73, 85–6, 289–90, 296–7
Consultants, in engineering 198

Consumer spending 133, 237–8, 239; cuts in 282–3; rising 86–7, 90
Contracting out of work 240
Cookson Group 241–2
Corby (BSC) 33, 43
Corfield, Sir Kenneth 147
Cornhill Insurance 99
"Corporatism" 12–13, 19, 156
Courage, brewers 13, 54, 215
Courtauld, Samuel 6
Courtaulds 13, 43, 54, 60–1, 238, 272, 275, 282
Coventry Hood and Sidescreen 51–2; as CH Industrials 117–18
Cowley 45, 60, 89, 188
Crane (John) & Co. 244
Creda 239, 243, 276
Credit 133
Cripps, Sir Stafford 9
Crown Paints 237, 250
Crown Wallcoverings 115
Crowther (John) & Co. 245
Cuckney, Sir John 207–8, 277
Cunard 258
Cunningham, Norman 120–1
Curry, Chris 143
Curtis, Alan 51

Daf 211, 230, 260, 262–3, 267
Dagenham (Ford) 270, 278
Daimler, Gottlieb 5
Dart Spring 48
Davies, Gareth 15, 28, 119, 162, 283
Dawson International 206–7
Day, Graham 141, 211, 258, 259, 261
Daylay 180
De Havilland Comet 10
De Lorean, John 79–80
De Vigier, William 120
Debuisser, Henri 87, 167
Decca 28, 146
Dedpan 51, 118
Dee Corporation 245
Deflationary budgets 44, 63
Delanair 279
Deming, Edwards 193
Denationalisation 258–9
Department of Industry 32, 220
Department of Trade and Industry 215, 292
Derecognition of unions 227
Derwent Valley Foods 290–1, 296–7
Derwentside Industrial Development Agency (DIDA) 289–90, 296
Design and development, speed required 302
Deutsche Bank 304
Dick, A. B. 153
Distillers Company Ltd 203, 207, 209, 212, 213

315

# INDEX

Dixon, Kenneth 254–5
Dollar, falling 286; rising 134, 136; weak 238
Duffus, Lyall 296
Duffus Clay 290, 296
Dunbee–Combex–Marx 43, 241
Dundee, Ford plans abandoned at 271
Dunlop 53, 121–30; and BTR 215–16, 241; and Sumitomo 231–2
Duport 249
Dyer, Henry 218

EETPU 220, 228–9, 234, 308
EMI 28, 147
Eagle Star 202
Economic policy 31, 271–2
Education, UK 4–5, 12, 308; Japan 156; Germany 5, 156
Edwardes, Sir Michael 25, 34–5, 41–2, 45, 51, 59, 60, 68, 69–70, 138, 169, 187, 189, 229; at Dunlop 127–30, 216; leaves BL 84–5
Edwardes, Roger 105
Edwards, Nicholas 35
Efficiency, lack in UK industry 10
Egan, Sir John 66, 69–70, 78–9, 198, 272–4, 286
Eglin, Roger 63
"Eikokubyo" (English disease) 219, 220
Elders IXL 202, 211, 281
Electrical, Electronic, Telecommunication and Plumbing Union (EETPU) 220, 228–9, 234, 308
Electronics industry 134, 238
Eliott, John 32, 202–3, 211, 214–15, 217
Employment *see* Jobs; Unemployment
Employment Act 1980 41
Employment Act 1982 86
Engineering Council 184–5
Engineering Employers Federation 284
Engineering industry 5–7, 47–8, 56, 238
Engineers 184–99
English Electric 13, 81, 152
English Estates 297
Enterprise agencies 293
Enterprise Allowance Scheme 89
"Enterprise culture" 290
Enterprise zones 54
Entrepreneurship 290–5
Eurofer 70
Euromarkets 72
Europe, as UK export market 9; trade rivalries 8
European Airbus 135
European Commission, study of UK industry 26; and BAe/Rover 262
European Community and Japan 221
Evans, John 47

Evered 108, 114, 243, 251
Exchange controls 33, 101
Export Credits Guarantee Department (ECGD) 137
Export markets, UK postwar 9
Exports 134–7; 1980 55–6; fall in UK, 1910s–30s 8

FKI 238
FMC 180
FT 30 index 44
Fairey Holdings 33, 138, 249, 259
Falklands factor 90
Far East, textile industry 135
Fashion industry 135–6
Faure 279
Fedden, Roy 5
Federal Reserve 44
Ferguson TV Group 147–8
Ferranti 24, 33, 52, 138, 214, 238, 244, 259, 277
Festival of Britain 8
Fiat 207
Finance for companies 303–4
Finniston, Sir Monty 15, 42, 184–5
Firestone 54
Fletcher, Leslie 59, 119
Flexible manufacturing systems (FMS) 82
Fodens 52
Fogarty 117
Fokker 267
Folkes, Constantine 46, 66–7, 242–3, 285–6
Folkes Group 46, 242–3
Ford, Henry 6–7, 185
Ford Motor Company 11, 59–60, 140, 188, 197–8, 216, 222, 225, 261, 278, 308; and Jaguar 286; labour relations 30, 68–9, 89, 269–71, 284; and Land Rover 210–11; professional management 17, 83–4
Ford Cosworth 253
Forte, Lord 157
*Fortune* 269
France, economic problems 90; postwar industry 8
Francis Industries 114
Franks, Julian 291
Fraser, Sir Campbell 121, 123–4, 125–6
Freight Rover 260
Fringe benefits 27–8
Fujitsu 75
Fulton 243

GKN 48, 57, 76, 114, 241, 256, 280
GPT 277
Garford-Lilley 113
Gateway 281

# INDEX

Gatward family 116
Gauntlet, Victor 117, 118
Geddes, Reay 121–2
Geddes Report 13
General Electric (US) 277–8
General Electric Co. (GEC) 15, 17, 28, 81, 145–6, 152–3, 170, 203, 207, 214, 267, 276, 277, 278
General Motors (GM) 253, 279, 286; Bedford Truck 209–11
General strike 1926 8
Germany, balance of trade 154; education for industry 5; and industry 4, 8, 42; state aid 53, 156
Ghafar bin Baba, Abdul 125
Gilchrist, John 22, 190, 230
Gill, Tony 11, 22, 23, 63, 67, 162, 163, 177
Gilmour, Sir Ian 42–3
Glaxo 76, 148, 238
Glynwed International 15, 28, 48–9, 65, 119, 283–4
Gold prices 65
Golden share, government, for Jaguar 139, 286
Golden Wonder crisps 209
Goldstar 298
Goldsmith, Sir James 96, 202, 281–2
Gooding, Terence 138
Goodman Fielder Wattie 281
Goodrich Tyre Co. 97
Goodwin, Phil 295
Government support, aerospace 10, 20; for Dunlop 123; German and US state aid 33; motor industry 11; shipbuilding 13; steel industry 12; *see also under the names of individual companies*
Graduates and industry 16, 17
Green, Sir Owen 97, 98, 102, 128–30, 158, 164, 170, 174, 182, 215–17, 304, 306
Grenier, David 44
Griffin, Arthur 48
Gripperrods 253
Gross national product 271
Guinness affair 200, 201–2, 207, 209, 211–13, 215
Gulliver, James 203, 207, 212

Halewood (Ford) 11, 60, 68, 89, 224, 270
Hallside (BSC) 35
Haloid 87
Hamilton, John 73
Hammond, Eric 228–9
Hammond, Roy 247
Hanson, James (Lord Hanson) 96, 97, 99, 102, 104, 157, 158, 208, 212, 215

Hanson Trust 17, 28, 95–6, 164, 170, 204, 208, 209, 212, 213, 244, 251, 308
Hanwell, Richard 111
Harland and Wolff 53, 266
Harrison, Bill 71
Harrison, Ernest 146
Harvey-Jones, Sir John 47, 126, 157, 158, 163, 164, 170–3, 175, 176, 212, 272
Hattersley, Roy 210
Hauser, Herman 143
Hayes, Martin 263
Hays, Ian 15, 227, 230, 232–3
Hayters 114
Healey, Denis 157
Hearley, Tim 51–2, 117–18, 214, 252–4
Heath, Edward 19–20, 20–1, 96
Heinemann 99
Henderson, Sir Denys 21, 22, 44, 58, 174, 175
Herbert, Alfred 6, 24, 32
Heseltine, Michael 207
Hewitt, Jeff 275
Hewlett Packard 234
High tech industries 42, 65, 82, 134, 142–53, 275
Hillsdown Holdings 108–9, 115, 164, 165, 166, 168, 180–1, 182, 201, 251, 257
Hinari 283
Hinchingbroke, Lord 7
Hirst, Hugo 152
Hitachi 152, 277
Hodgson, Sir Maurice 58, 126, 127, 129, 170, 171
Hoechst 250
Hogg, Sir Christopher 60–1, 66, 282
Holdsworth, Trevor 57
Honda 27, 34, 76, 188, 210, 267–8, 278, 300
Hoover 54, 283
Hornby 43, 241
Horrocks, Ray 25, 76, 138, 210
Hotpoint 49, 239, 276
Howard Johnson 203
Howe, Sir Geoffrey 33, 44, 56, 61, 62
Hutchings, Greg 95, 114, 247, 251
Hutchinson, Eddie 296

IBM 64, 142, 143, 187, 233–4
ICL 13, 33, 53, 57, 61, 64, 127, 138, 139
IMP *see* Integrated Micro Products
I'Anson, Mark 73, 85–6, 151–2, 290, 293
Imperial Chemical Industries (ICI) 7, 13, 43–4, 44, 57, 84, 87, 170–3, 174, 175, 176, 187, 238, 244, 250, 256, 285, 309; and the government 5; management style 233; problems 58, 285
Imperial Foods 109
Imperial Group 181, 203–4, 208, 209, 212

317

# INDEX

Imperial Tobacco 13
Imperial Typewriter Co. 10–11
*In Place of Strife* 16
Ind Coope 54
Industrial and Commerical Finance Corporation (ICFC) 85; *see also* Three-I
Industrial relations 16; *see also* Strikes
Industrial Relations Act 20
Industrial Reorganisation Corporation (IRC) 13, 19, 24, 97, 152, 247
Industry, decline in UK 3–4, 8–9; British attitude towards 4–7; Conservative party policy ix, x; status quo 15–16; structure post 1980 recession 65
Industry Act 1972 20
Industry Year 1986 200
Inflation, 1970s 20–1, 23–4, 28, 32, 34; 1980s 43, 46, 238, 271, 272, 280; down 53, 57, 78, 90, 238
Information technology (IT) 63, 82, 144–5
Ingersoll Engineers 186, 191–2, 198
Inmos 24, 33, 53, 76, 139, 144, 147–8, 259, 276
Integrated Micro Products (IMP) 85–6, 151–2, 290, 297
Interest rates, up 33, 44, 59, 239, 271–2, 278, 282–3; down 63, 78, 239, 271
International Monetary Fund 21, 28
International Signal and Control 244
Interventionism 20, 24, 244–57; anti- 33; constructive 53, 79–80
Invest in Britain bureau 220
Investment 11, 274
Investment management 100
Investors in Industry *see* Three-I
Ireland, Norman 97
Iron and Steel Trades Confederation (ISTC) 39–41
Ishihara, Takashi 221
Issigonis, Sir Alec 14, 35
Istel 260
Isuzu 209, 279

JCB 56
Jackson, J. and H. B. 113, 249
Jackson, John 115
Jackson, Tom 34
Jaeger 205
Jaguar Cars 25, 69–70, 78–9, 83, 134, 136, 197–9, 225, 272–3, 286; flotation 139, 259
Jamieson, Bill 106
Japan, general 4, 8, 300–1, 305–6; economy 136, 173, 221, 274; and Europe vii, 221; management 68–9, 79, 83, 198–9, 222–7, 229–30, 231–2, 264, 269, 270, 298; quality control 183, 193–4; and unions 231–2; manufacturing methods: flexible systems 188; robots 190–1; products: narrow ranges 186; microchips 143–4; microwaves 26; motorcycles 26–7; rubber 124–5; steel 14; in the UK 76, 240, 278–9; investment in Britain 71–2, 124–5, 217, 218–20
Jay, John 106
Jenkin, Patrick 80–1, 137
Jenkins, Simon 309
Jobs: creation and training 61; losses 65, 66; security 13, 15, 223; vacancies 134–5; *see also* Redundancies; Unemployment
Jones, Barry 47
Jordan, Bill 281, 284
Jordan, Michael 121
Joseph, Sir Keith 3, 31–3, 34–5, 40, 42, 44, 46, 47, 52, 53, 54, 56, 57, 59, 60, 65, 66, 69, 70, 72, 75, 85, 123, 144, 165, 291
Junk bonds 202
Juran, Joseph 193
Just-in-time (Kanban) 195–7, 226, 299

"Kaizen" 194
Kaletsky, Anatole 54
Kanban (just-in-time) 195–7, 226, 299
Kearton, Sir Frank 13
Keith, Lord, of Castleacre 147, 162
Kent, Geoffrey 203–4
Keynesian economics 31
Kidde 244, 251
Kiep, Walther 176
Kilroot (ICI) 44
Kinder, John 117
King, Lord 157
King, Tom 134
Kirkby Co-op 24
Kleinwort Benson 112, 207
Komata, Satashi 223–4
Komatsu (UK) 218, 221–6, 279, 299
Komiya, Torio 226
Kraske, Karl-Heinz 277

LEK Consultants 252
Labour Party/Government, and the economy 239; and industry 12–13; and internationalisation 176; and nationalisation 259
Lacey, Graham Ferguson 50–1, 75, 213
Laidlaw, Christopher 64, 75
Laing, Sir Hector 203–4, 208, 209, 212, 213
Laird Group 210
Laissez-faire economy 7
Laister, Peter 147–8
Lake and Elliott 114
Lame ducks 24, 138–41
Land Rover 209–11, 260, 302
Langston, David 296
Lawson, Nigel 14, 139, 155, 271–2, 273, 278

# INDEX

Leach, Rodney 258
Lesney 43
Levene, Peter 145–6, 214
Lever, Paul 250
Lever, William 6, 98
Lever, Lord 40
Leveraged buy-out (LBO) 175–6, 202
Lewinton, Chris 243
Lewis, Sir Edward 146
Leyland and Birmingham Rubber Co. 97
Leyland Bus 211, 260
Leyland Daf 262–3, 267
Leyland Trucks 189–90, 226, 260, 262–3
Leyland Vehicles 14, 16, 65, 69, 209–11, 230
Ley's Foundries and Engineering 104, 107, 249
Liddell, David 73, 85–6, 293
Linwood 11, 62, 83
Litton Industries 10–11
Llanelli Radiators 260
Llanwern 71
Loan Guarantee scheme 72
Lockwoods 109, 180
London and Midland Industrials 249
London and Northern 104, 251
London Brick 204, 283
London Chamber of Commerce 227
London Enterprise Agency 293
Longbridge 45, 60, 68, 76
Lonrho 50, 96, 211
Lord, Alan 123, 124, 125, 127, 163
Lucas Industries 11, 22, 48, 55, 57, 63, 65, 67, 177–8, 199, 241, 256, 274, 280, 308
Lygo, Sir Raymond 261, 262

MFI 283, 294
MG 51–2, 84
MRP system 190, 196, 226
McAdam, Jim 204, 205–7
McGhee, Ray 296
McGowan, Brian 104–7, 113, 181, 249, 250, 251, 282
MacGregor, Sir Ian 41, 61; at BSC 47, 70–1, 80, 88–9, 137, 162, 265
Machine tool industry 24, 274
McKechnie, Roger 296
McKechnie (company) 249
McKinsey 17, 124, 170
Macmillan, Harold 10, 12
Magnin, Roland 87–8, 183
Magowan, Lord 171, 233
Malaysian Rubber 125–6
Management 27–8, 32, 161–2, 167–8; *see also* Japan: management
*Management Today* 17
Managers 55, 66
Manpower (company) 244

Manufacturing *see* Industry
Manufacturing Resources Planning (MRP2) 190, 196, 226
Marconi 134
Market segmentation 17
Marks and Spencer 83, 197, 206
Marsh, Roy 122, 125, 127
Massey Ferguson 54
Matchbox Toys 43
Mather and Platt 238
Matrix management 170
Maxwell, Robert 111–12, 150, 163
Meaney, Sir Patrick 102, 103
Meccano 35–6, 43
Meggitt 115
Melchett, Lord 15
Merchant banks 100
Mergers 11, 13–14, 200–1; ICI 176; motorcycle industry 26; *see also* Take overs
Meriden Motorcycle Co-operative 24, 52–3
Mersey Docks and Harbour Board 20
Merseyside 11
Messervy, Godfrey 177
Metal Box 40, 61, 65
Metro-Cammell 210, 214
Metsun 277
Michelin 122
Microchip industry 24, 143–4
Microprocessors 73
Microwaves 26
Midland Bank 80, 128
Miller, Ronald 206–7
Mills, Gavin 112
Miners' strike 1972 20; 1984 133
Ministry of Defence contracts 145, 153, 259, 263
Mitcheldean 87
Mitel 75
Monetarism, origins 21
Monopolies 11, 32
Monopolies and Mergers Board/Commission 153, 203, 208, 214, 255, 275
Monroe Auto Equipment 279
Montague, Michael 244
Montupet 279
Morgan Grenfell 211–12, 213
Morris, Barry 16, 227, 230
Morris, Gavin 247
Morris, Sir William, Lord Nuffield 5, 6
Morris Motor Co. 14
Motor industry, Germany 5, 10; UK 5, 8, 9–10, 11, 22, 48, 65, 238, 278–9; mergers 13–14; US 10; *see also under names of individual companies*
Motor Panels 253
Motorcycle industry 26–7

319

# INDEX

Moulton, John 85
Mountfield 299
Multinationals, sales by 16–17
Murdoch, Rupert 163
Murphy, Sir Leslie 24

NCC Energy 75
NSK 285
Nader, Ralph 52
National Coal Board 133
National Economic Development Council (NEDC) 13
National Enterprise Board 24, 25, 32, 33, 52, 72, 138, 149, 259; abolished 74
National Institute of Social and Economic Research 46, 307
National Semiconductor 144
National Westminster Bank 127, 247
Nationalisation, aerospace industry 24; denationalisation 258–9; shipbuilding 24; steel industry 12, 14–15
Nestlé 255–6, 257
Newman Tonks 249
News International 227
Newton, David 181
Nicholson, David 258
Nigeria 69, 137
Nimrod reconnaissance plane 145–6, 214, 276
1992 and takeovers 254–7, 304
Nissan 188, 219, 220, 298; in Britain 71, 199, 221, 222, 223, 224–5, 270, 278, 279, 280
Niven, Keith 303
Norcros 249
North-East, and Japan 222
North-East Shipbuilders 266
North Sea Oil 28–9, 59, 63
North/South divide 62, 138, 283
Northern Ireland, unemployment 62
Norton Opax 111
Norton Villiers Triumph 27
Norwich Union 111–12
Nottingham Manufacturing 206
NuTone 244

OPEC 34, 137
Office of Fair Trading 153, 208, 255, 275
Oil Feed Engineering 97
Oil industry, crises 20–1, 28, 29; *see also* North Sea Oil
Oil prices, falling 79, 137, 239; rising 34
Oliver Wight Organisation 190
Olivetti and Acorn 143
Olsen (Fred) Co. 266
OPEC 34, 137
Opel 140

PA Consultants 198, 264
Paint manufacturing 250
Palmer, Brian 283
Parker, Sir Peter 162
Parkinson, Cecil 138
Parkray 243
Parnaby, John 177, 199
Parnell and Sons 214
Pay rises, 1980s 46
Pay policies 20–2
Pay restraint policy 28, 30
Peacock, Alan 53
Pearson, David 299, 300
Pegi Malaysia 125, 127, 130
Pendragon 282
Pennock, Sir Raymond 59
Pension fund investments 100, 101–2
Perkins Engines 307
Perry, David 111, 112
Peters, Tom 170
"Petrocurrency" 29
Peugeot 62
Peugeot Talbot 83, 140, 280
Pharmaceutical industry 7, 148–9, 238, 280
Philips 56, 277
Photocopiers 87–8, 183
Pickens, T. Boone 202
Pilgrim House 248, 251
Pilkington 216–17, 293
Pirelli 121–3
Plastow, Sir David 179
Plaza Accord 1985 136, 300
Plessey 28, 145, 146, 152–3, 203, 214, 238, 276, 277–8
Polycell 250
Port Talbot 71
Post Office 146, 153
Povey, Phil 55
Powell Duffryn 204
Predators 95–109, 290–1; *see also* Take-overs
Prestcold 107
Pretty Polly 99
Price (C.) and Co. 104, 105
Price Commission 20, 21, 22, 33
Price/earnings ratio 106
Price Waterhouse 122
Prior, James 41, 214
Privatisation 138–9, 258–68; of British Shipbuilders 265–6; of British Steel Corporation 265; of British Telecom 258
Product ranges 186
Production, forms of 185
Productivity 154, 186–7
Professional and Executive Register 55
Profits 56, 57, 239–40

# INDEX

Protection of industry, Japanese 156; against Japan 136, 219; UK 7–8
Prudential 99, 103
Ptarmigan Telecommunications System 153
Public sector 43
Public spending cuts 33, 63
PurePlas 296
Purser, Christopher 49

Quality, in production 193, 198
Quality circles 69, 197, 226, 230, 270
Quantec Systems and Software 88

RFD 50, 213
RHP 13, 247–8, 251, 285
RTR 28
Racal 28, 145, 146–7, 214, 275–6
Radford, Gerry 231
Radio and Allied Industries 152
Raleigh Bicycles 243
Rank Hovis McDougall 256, 281
Rank Xerox 87–8, 183, 185, 198, 240, 302
Ransomes, Hoffman and Pollard (RHP) *see* RHP
Ravenscraig 80–1, 88–9, 137, 265
Rawlplug 181–2, 250
Recession 1980 42–4, 46–61; aftermath 62–77; 1982 79–82
Redundancies 54, 55, 283; British Leyland 25; and Conservative government 44; engineering and motor industries 48
Reed Paper 54, 250
Regional aid, reduction 32, 33
Regional policy, UK 11, 54, 138
Regional variations, in jobs 133–4
Renault 262
Renold 48
Rent-a-Center 276
Research 10, 12
Restrictive practices, legislation 11
Revlon 54
Ridley, Nicholas 286
Risk, Sir Thomas 207, 213, 215
Robinson, Bernard vii, 82–4, 163, 164, 298, 299, 303, 304, 309; and automation 193–7, 199, 301
Robinson, Derek 23, 35, 41
Robinson (Thomas) and Co. 245
Robots 82, 84, 188, 190–1, 301
Roderick, Ian 88–9
Rodgers, Tony 84, 175, 233
Rodime 142, 275
Rolls-Royce 20, 32, 60, 64, 139, 185, 187, 302; automation 191–2, 198; privatisation 259, 260; strikes 284
Rolls-Royce Motor Cars 57, 81, 178–80

Rootes Group 11
Routledge, Jim 86
Roux, Olivier 202, 213
Rover Group 211, 259–62, 267, 306
Rowland, Tiny 163
Rowntree's 49, 254–6, 257
Royal Enfield 263–4
Royal Ordnance 259, 260, 263–4
Rudd, Graham 245
Rudd, Nigel 95, 104–7, 112, 164–7, 181, 182–3, 237, 249, 250, 251, 257, 282
Russell Hobbs 243
Ryder, Donald 25

SGS 276
SKF 186, 247, 248
SPC *see* Statistical process control
SP Tyres 231
STC 139, 145, 146, 147, 277
Saba, Shoichi 176
Sampson, Anthony 129
Samsung 298
Sanderson, Roy 220, 222
Saunders, Ernest 17, 201–2, 207, 209, 212–13
Scargill, Arthur 133, 228
Scholey, Bob 70, 265
Schroder's 113, 217, 303
Scientific Management (Taylorism) 6
Scott Lithgow 141
Scottish and National Breweries 281
Scottish Development Agency (SDA) 142
Scunthorpe (British Steel) 71
Securities and Exchange Commission, US 215
Security of employment 223
Seelig, Roger 212
Serck 99
Sergeant, Patrick 103
Service industries ix, 133–4, 155
Servis (company) 285
Set-up times 301–2
7-eleven 202
Shioji, Ichiro 221
Shipyards 10, 23, 43, 44; and Government 13, 24; job losses 65
Short Brothers 266–7
Short-termism 302
Shotton (British Steel) 33, 35, 43
Siebe 114
Siegel and Stockman 253
Siemens 276, 277, 278
Sikorsky 207, 209
"Silicon Glen" 143, 144
Silkin, John 47
Simon, John 126
Simpson, Duncan 55
Sinclair, Sir Clive 24, 63, 82, 143, 148–51

321

# INDEX

Sinclair Research 148, 149
Singer 49
Single European Market 1992 256, 304
Sirs, Bill 39–41, 60
600 Group 82
Skill shortages 134
Slater, Jim 13, 20, 95, 96, 97
Slater, Walker 14
Sloss, Ian 124, 231–2
Small Engineering Firms Investment Scheme 89
Small firms 18, 292–3
Smallbone 250
Smedley's 180
Smith, Roland 260–1, 262
Smith and Wesson 251
Sobell, Michael 152
Social contract 21
Solomon, Harry 108–9, 115, 157, 163, 165, 166–7, 168, 180, 251
Sony 149, 219, 221
South Africa 119
South African subsidiaries 65
Southgate, Colin 276
Spectrum Computers 149–50
Spirella 206
Spurrier, Henry 9
Staffordshire Potteries 117
Standard Chartered 247
Standard-Triumph 14
State aid, Germany and US 33
Statistical process control (SPC) 194, 230, 232
Stauffer Chemicals 244
Steel industry 12, 14–15, 46
Stephens, Barrie 114
Stephenson, Keith 293, 295
Sterling, down 78; high 66, 271; strength, and exports 155
Stock Exchange regulation 211
Stock market, bullish 91, 99, 133, 201; crash October 1987 237; French and German 100; reform 100–1
Stocksbridge 71
Stokes, Donald 9, 14, 25
Stone-Platt 121, 128
Stop-go policy on credit 26
Storeys Industrial Products 75, 85, 89, 118
Strathclyde Institute 299
Strikes 22, 23, 30, 34, 39–41, 280–1; 1987 226; no-strike packages 220; and productivity 11–12
Suchard 254–6
Sugar, Alan 150–1, 163, 290
Suggestion schemes 230, 233, 300
Sumitomo Rubber Industries 124–5, 231–2
Summerfield, Peter 263–4

Suter 107–8, 114, 251
Suzuki 27, 209
Swish (company) 250

TI Group 114, 243–4
Take-overs 99–100, 101, 238, 244–5, 281–2; see also Mergers; Predators
Talbot Engines 54, 55
Tallent Engineering vii, 82–4, 193–9, 222, 241, 279, 298, 299, 301
Tapley, Don 97
Tax measures 33, 63
Taylor, Lord 157
Taylor, Brian 49–51, 75, 85, 89, 118, 164, 165, 213, 252, 294, 302
Taylor, Frederick 6
Tebbit, Norman 70, 86, 138, 139, 157, 209
Television sets 26, 149
Telford 48
Temporary Employment Subsidy 21–2
Textile industry 23, 43, 60–1, 135–6, 238, 272; job losses 65, 275
Thatcher, Margaret 3, 53, 59, 116, 136, 157; and British Leyland 34–5, 210–11; re-elected 1982 90; 1987 238; on unions 34
Thatcherism, seeds of 16, 18, 63
Third World competition 60
Thompson, David 108–9, 115, 165, 166, 180, 251
Thomson (French company) 276, 277
Thomson, J. Walter 201, 244
Thorn-EMI 139, 144, 145, 147–8, 276
Thorn Electronics 26, 54
Three-I (3I) 85, 246, 272, 291–2, 293, 294, 295

Tikkoo, Ravi 266
Tilling, Thomas 90–1, 99, 102, 103, 217
Tillotson, Oswald 96
Timex 149
Tioxide 241–2
Tolley, Leslie 42
Tomkins, F. H. 114, 251
Tootal 275
Toshiba, in Britain 271
Toshiba/Rank 219, 220
Total quality control x, 193–4, 198, 226, 230, 232, 299
Toymaking industry 43
Toyota 173, 188, 195–6, 278–9, 301
Trade, world 10, 280
Trade barriers in overseas markets 80; see also Protection
Trade deficit 1987 239; 1989 278
Trade figures 1980 55
Trade gap 1987 271–2
Trade Union Act 1984 137

# INDEX

Trade unions x, 8, 16, 22, 23, 68, 227–8; American union busters 227; and British Steel 60; and Conservative government 20–1, 137; and Ford 68–9, 89, 284; and mergers 13; and new techniques 195, 197, 228–30; and pay restraint 30; and redundancies 55; and Social Contract 21
Trades Union Congress 16, 20, 34, 41, 86, 229; Day of Action 46
Trafalgar House 141, 258
Training, lack, in UK 306–8
Training and Enterprise Councils (TECs) 307
Transport and General Workers Union (TGWU) 45, 228
Trebor 136
Trippier, David 293
Triumph Motors 11
Trusthouse Forte 212
Tube Investment (TI) 17
Tudor Webasto 117, 214, 253
Turnbull, George 140
Turner and Newall 75
Tycoons, personalities 163–5

Unemployment, falling 1980s 280; rising 1970s 21, 31; rising 1980s 43, 54, 62, 65, 66, 90, 240, 271; Northern region 47
Unions see Trade unions
Unipart (UGC) 178, 260
Uniroyal 121
United Biscuits 203–4, 208, 209, 211, 212
United Drapery Stores 104, 204
United States, aid to UK 9; economy 79, 90, 134, 136, 239; industrial aristocracy 4; industry postwar 8; methods of management in UK 232–4; ownership of UK companies 187; productivity 189–90; role model? 306
Universities, cuts in funding 74; and industry 4–5, 12
Unlisted securities market 294
Utiger, Ronny 243

VSEL Consortium 258, 266
Valentines 110
Valeo 279
Valor 244, 245
Valor Bruce 214
Value added tax (VAT) 33, 34, 56
Vantona Viyella 135–6, 206
Vauxhall 48, 140
Venture capitalism 293–5
Vickers 57, 81, 179, 258
Videomaster 110
Villiers, Sir Charles 47, 52

Volcker, Paul 44, 176
Volkswagen 53, 261
Volvo 211, 260

WETS 114, 249, 251
WPP 244
Waddington (John) and Co. 110–12, 163, 241
Wage restraint 42
Wages 240, 280
Walker, David 302
Walker, Peter 14
Wallbridge 117
Wallmates 116
Walls, Stephen 277
Wang 142, 275
Warburg's 130
Ward, Peter 178–80
Ward White 49
Wardle, Bernard 49, 50, 75, 85, 213
Wardle Storeys 118, 213, 252, 279, 285
Warwick University 12, 16, 189
Washing machine trade 26
Washington, Co. Durham 221, 224
Waterman, Robert 170
Watson, Victor 110–12
Weinstock, Arnold, Lord Weinstock 13, 81, 145–6, 152–3, 155–6, 158, 174, 204, 209, 214, 276–7, 278, 306
Welfare state 9
Welsh Development Agency 105
West Midlands 74, 138
Westland 200, 207–9
Wharton, Les 209
White, Sir Gordon 96, 97, 244
White, John 115
White, Philip 113
Wickens, Peter 224
Wiener, Martin 4
Wiles Group 96
Williams, W. and Sons 104, 105, 106, 107
Williams Holdings 107, 108, 113, 114, 138, 164, 167, 168, 181, 182, 216–17, 237, 248, 249, 250, 251, 257, 282, 283
Willis, Norman 271
Wills, Nicholas 246
Wills Fabrics 136
Wilmot, Robb 64, 75, 146
Wilmot Breeden 48
Wilson, Harold, Lord Wilson of Rievaulx 12, 19, 21, 35, 162
Wilson Commission 72
"Winter of Discontent" 30, 43
Worboys, Sir Walter 97
World economy 28, 29
Wyman, Tom 176

323

# MANUFACTURING IN THE 1980s

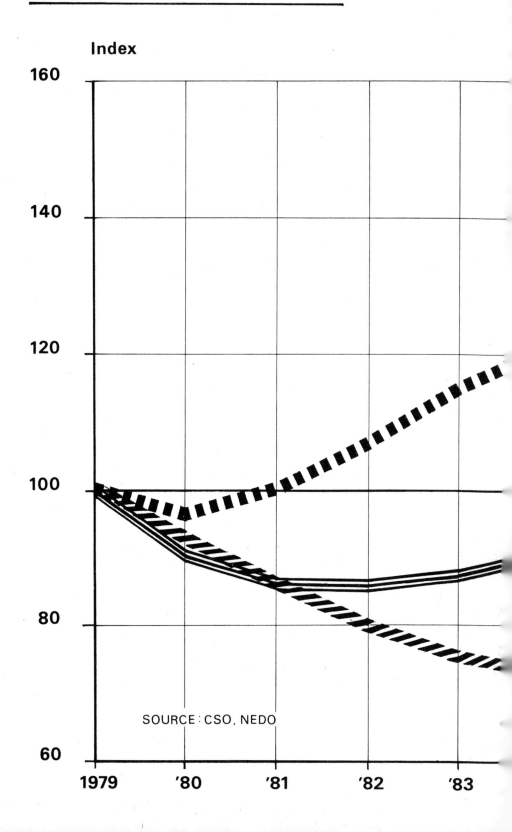